Diagnosis

Diagnosis

SOLVING THE MOST BAFFLING MEDICAL MYSTERIES

Lisa Sanders, M.D.

B\D\W\Y
BROADWAY BOOKS
NEW YORK

Published in the United States by Broadway Books, an imprint of
Random House, a division of Penguin Random House LLC, New York.

crownpublishing.com

Broadway Books and its logo, B \ D \ W \ Y, are trademarks of
Penguin Random House LLC.

The columns in this book originally appeared in
The New York Times Magazine.

Library of Congress Cataloging-in-Publication Data
is available upon request.

ISBN 978-0-593-13663-8
Ebook ISBN 978-0-593-13664-5

Printed in the United States of America

2 4 6 8 9 7 5 3 1

First Edition

Book design by Susan Turner

To the patients who have shared their stories with me—in my office, in my column, and in this book

CONTENTS

INTRODUCTION

Solving the Puzzle

The lights in the doctor's office were almost too bright for the fifty-year-old woman to bear, but she forced herself to open her eyes. A young doctor knocked, entered the exam room, and introduced herself. She seemed sympathetic and interested as her patient described her miserable week and the travel that preceded it.

She hadn't felt well since she and her kids had gotten back to Chicago from their two-week trip to her parents' home in Kenya. This had been her first visit in nearly a decade—since before her kids were born. And now that they were old enough, she had been eager to show them where she grew up. She had gotten them all the right vaccines, and she had made sure they took the medicine to prevent malaria every single day. She didn't want the trip, or their memories of this place she loved so much, to be spoiled by illness. It was a great trip.

But coming home had been brutal. Her children recovered after a day or two of jet lag. She didn't.

She gave it a week, but every day she felt worse. She was tired, as if she hadn't slept for weeks. She was nauseous. She felt hot and sweaty, like she had a fever. And her body ached like she had the flu. She called her doctor's office, but her doctor was out of town. So she found another who, miraculously, was able to see her the following day. And now here she was.

The patient paused, then added, "I think I've felt like this before." When she was seven years old, living in Kenya, she had a bout of malaria. She thought maybe that's what she had now. It sure felt like it.

The doctor nodded—it was a reasonable theory. Malaria is endemic in regions like sub-Saharan Africa and the most common cause of fever in travelers returning from there. And since she'd had it before, she knew the achy, flu-like symptoms of the blood-loving parasite.

Still, the doctor told her, she'd need a little more information. Any other medical problems? Absolutely not. Before her trip, she'd been completely healthy. She took no medications. Didn't smoke or drink. Worked in an office. She was divorced and lived with her two children. She'd taken the preventive meds every day, starting two weeks before their trip, as prescribed.

The doctor moved the patient to the exam table. She didn't have a fever, but she'd taken acetaminophen earlier that day. She was a little sweaty, and her heart was racing, but otherwise her exam was unremarkable.

Malaria made sense to the doctor. There is a type of malaria in parts of Kenya that isn't killed by the usual prophylactic medications. And if she had already had the infection for more than a week, it was important to start treatment quickly, the doctor told her. She gave the patient a prescription for a three-day course of antiparasitic medications. The woman took the prescription gratefully. She looked forward to feeling better at last.

This is the usual story of diagnosis. A patient feels sick. She recognizes that there is something wrong, but she may wait a day or two before seeking help. Things often get better on their own. But when they don't, she will often seek help from her doctor.

From there, it's the doctor's job to solve the puzzle. Listening to the patient's story is key. In nearly 80 percent of cases,* that is where the most important clues can be found. An examination may offer additional clues. Sometimes a test reveals even more. And it's up to the doctor to put it all together and make the diagnosis.

Before I went to medical school, all I knew about diagnosis was what I'd seen on TV. It was an almost instantaneous one-liner dropped at a dramatic moment—just after the patient's opening story of symptoms and suffering and just before they're whisked away for a life-saving treatment. I believed that diagnosis was a puzzle that I, once I was a doctor, could easily solve.

During med school, I put in the hours to learn the building blocks of diagnosis—chemistry and organic chemistry, physics, physiology, pathology, and pathophysiology. As I finished my schooling and started on the apprenticeship component of my training, I developed a series of what doctors call "illness scripts"—detailed inventories of symptoms

* Hampton JR, Harrison MJG, Mitchell JRA, Prichard JS, Seymour C, *British Medical Journal*, 1975, 2, 486–89.

and their variations, progressions, and resolutions, which create a picture of a particular disease. Once these scenarios were committed to memory and mastered, they could be deployed as needed. Nausea, vomiting, and diarrhea that rapidly sweeps through a family is a viral gastroenteritis. The sudden onset of fever, body aches, and congestion during flu season means the flu. Or in this case, those same symptoms in a traveler returning from Kenya likely mean malaria. We see the symptoms. We recognize the pattern and immediately know the diagnosis.

Fortunately, that is what happens most of the time—up to 95 percent of the time, according to one study.* It is a skill set that delivers—most of the time. But what about those other cases? The 5 percent where the doctor has no answer. Or worse, the wrong answer?

The sick woman thought she had malaria. So did her doctor. But after the three days of pills, she felt even worse. She was so weak she could hardly move. She vomited nonstop. She felt feverish. Sweaty. Her heart pounded furiously. She couldn't eat for four days and couldn't even get out of bed for two. Finally, she called the doctor, who promptly sent the woman to the emergency room.

In the ER, an examination showed the woman's heart was racing and her blood pressure was high. Her white-blood-cell count was dangerously low, and her liver showed evidence of injury. It wasn't clear what was wrong with her, so she was admitted to the hospital.

The doctors in the hospital gave the woman medicine to stop the vomiting. That helped. But after several days it still wasn't clear what had made her so sick in the first place. It

* Singh H, Meyer AND, Thomas EJ. *BMJ Qual Saf* 2014; 23:727–31.

clearly wasn't malaria. She'd had three blood smears examined in the lab. And although she wasn't running a fever when the blood was drawn—that's when the test for malaria works best—none of these smears showed any sign of the parasite that causes this potentially deadly illness.

Her doctors hypothesized that her symptoms were a reaction to the medications she'd been prescribed for the malaria they now knew she didn't have. That seemed possible, especially since she was feeling a little better. Once she could eat, she was discharged from the hospital.

But back at home, the patient started throwing up again. She toughed it out for a week but finally dragged herself back to that same community hospital. The doctors there were worried enough by her condition to transfer her to Rush University Medical Center, where many of them had been trained. They were certain that their colleagues at Rush would be able to solve the puzzle.

The doctors at Rush consulted an infectious disease specialist—what else could this woman have? She was in the hospital for a week. She saw so many doctors. She had so many tests. When the vomiting was under control and she could eat, she was sent home and told to follow up as an outpatient with the infectious disease doctor. But after a few days she was back at Rush, just as sick as she'd been the first time.

More doctors, more tests. Tests of her urine, her stools, her blood. CT scans, MRIs. Even a biopsy of her liver. The results weren't all normal, but they also didn't seem to add up to a clear diagnosis. She was given half a dozen antibiotic, antiviral, and antiparasitic medications. If the doctors couldn't figure out what she had, they could at least try treating her for what she might have. But none of the medications helped. What could she have picked up in Kenya? The dozens of doctors she saw were all asking that same question.

This is, perhaps, the most uncomfortable place to be in medicine, the land of uncertainty. It is uncomfortable for the patient—because not only are they still suffering from the symptoms that led them to seek care, they still don't know the cause. Will it get better on its own? It hasn't so far. Isn't there a test for this? And yet dozens of tests, sometimes more, have been unrevealing. Will it kill them? How can anyone give a prognosis without a diagnosis?

It's also uncomfortable territory for the doctor. One of the reasons that it may take doctors several tries before they discover the right diagnosis is that uncommon diseases often look a lot like their more run-of-the-mill counterparts early on. The body has only a few basic ways to let us know something's wrong—what we call symptoms. But the possible causes of those symptoms are many. It's like the relationship between letters and words—only twenty-six letters, but millions upon millions of words. In medicine there are dozens of symptoms. But according to the International Classification of Diseases, there are nearly ninety thousand diagnoses.

Of course, no single doctor knows all ninety thousand—though some doctors know a lot more than others. Once the possibility of an unusual diagnosis is raised, there are a few ways to supplement the knowledge a doctor lacks. There's the old-fashioned but often effective method of simply asking a colleague. Or there's the much newer method of consulting the peripheral brain—the Internet.

But even when we have all the data, a condition can still go undiagnosed. A disease recorded on a page or in a database often looks very different from the same disease living in a patient. The earliest studies of diagnosis, done in the 1970s, showed that the doctor most likely to make a difficult diagnosis was the one who had seen the disease before, firsthand. Personal experience can be more important than book knowledge.

After several weeks in and out of the hospital, the patient found herself at home, too weak and too sick to care for her children. She called her closest friend and asked her to stay with her and her children while she tried to recover. "Of course," the friend told her, and quickly packed a bag. Upon arriving at the woman's apartment, the friend was horrified by the woman's appearance. Her face was thin and gray. Her lips were pale. "You have to call your doctor," the friend said as soon as she heard the woman's story. "Dr. Brown will know what to do."

Dr. Marie T. Brown had been the woman's physician for more than twenty years. The woman called the office and made an appointment later that week. Dr. Brown was also shocked by the appearance of this patient she knew so well. Normally she'd see her once a year for a routine physical. They'd catch up on life and health and then say goodbye until the next year. She always looked healthy and robust. But not now.

When Dr. Brown entered the exam room, the patient was hunched over a basin, and the acrid scent of vomit filled the air. She had clearly lost a lot of weight, and her eyes and cheekbones were prominent on her much thinner face. Her left leg trembled and jerked uncontrollably. What in the world happened to you? the doctor asked.

The patient, with the help of her friend, recounted the events of the past weeks. Dr. Brown didn't have access to any of the patient's hospital records, so she knew only what the patient could tell her: that she hadn't felt well since returning from Kenya; that the doctors initially thought she might have malaria, but now they weren't sure what she had. And that she had never felt so sick and weak in her life.

Could she get on the exam table? Dr. Brown asked the patient. The doctor and the friend helped her up.

Starting at the head, Dr. Brown worked her way systematically down the woman's body. At the neck, she stopped. The patient's thyroid gland was much larger than normal. It wasn't tender, but it was big. Dr. Brown was pretty sure that was new.

She finished the exam quickly. The patient's reflexes were wild. A little tap sent arms and legs flying. And the left leg seemed to have a life of its own: shaking, jerking, trembling. She excused herself and stepped out, "to read up on something."

When Dr. Brown returned a few minutes later, she was pretty certain of the diagnosis. The patient had hyperthyroidism. She might even be in thyroid storm, the most severe form of the disease. Everything fit according to the conventional script—the racing heart; the sweatiness, shakiness, itchiness, and occasional fever; the weight loss. Everything except the vomiting. She had excused herself in order to make certain that vomiting could be a part of a hyperthyroidism diagnosis. She discovered that it was an unusual symptom, but one that has been seen before in others with hyperthyroidism. By late afternoon the diagnosis was confirmed, and Dr. Brown made arrangements for the patient to see an endocrinologist right away.

Seeing the solution makes it possible to recognize how it

might have been missed. Clearly, the patient's own instinct that her illness had started during her trip home played a role. Her interpretation of her symptoms—that she was febrile and that she felt the way she had when she contracted malaria forty years earlier—led the hospital doctors down a path to the wrong diagnosis. The onus doesn't fall solely on the patient, though. Once the doctors recognized that it wasn't malaria, they continued to limit their possible diagnoses to infectious causes.

None of the hospital doctors focused on her thyroid. Maybe they didn't see it? Doctors miss more by not seeing than not knowing, according to William Osler (1849–1919), the philosopher king of early internal medicine. On the other hand, goiters—the name we give to enlarged thyroids—are uncommon in the United States but very common in iodine-deficient areas like sub-Saharan Africa. According to the World Health Organization, over one quarter of children growing up in Africa will develop a goiter.* And once a thyroid is enlarged, it generally stays that way. So a goiter seen in a woman who grew up in Kenya may not seem remarkable to the average physician. The patient's own doctor, however, immediately recognized that the enlarged thyroid was new.

These are the cases, the ones that elude immediate recognition and diagnosis, that can be the most terrifying. But they can also be the most riveting and revealing. They illuminate how doctors think about patients and apply their knowledge,

* Andersson M, Takkouche B, Egli I, Allen HE, de Benoist B: Iodine Status Worldwide, WHO Global Database.

and they demonstrate how doctors and patients can work together to answer the patient's essential question: "What's wrong with me?"

These are the cases I write about in my *New York Times Magazine* column, Diagnosis, presented here in this collection. Each and every one is a detective story. Here the stakes are high and the risks great. Here the doctor must don her deerstalker cap and try to solve the mystery before her. Seeing these cases unfold reveals the difficulty of delivering a diagnosis that strays off script and defies the list of usual suspects. It also exposes the flaws in the systems that guide our medical practices, flaws that become visible only when the machinery is stressed.

I've organized the following chapters by symptoms—eight of the most common problems that send a patient to the doctor's office or the emergency room. While each story in a section begins with the same basic symptom—a fever, a splitting headache, a bout of nausea—all of them almost immediately take off in their own unexpected direction. So few symptoms, so many diagnoses.

In this book I try to put you, the reader, in the doctor's place. I want you to see what the doctor sees. I want you to feel the uncertainty of a puzzling disorder—and the thrill when that puzzle is solved.

PART I

Burning with Fever

Just a Fever

"I think I'm losing this battle," the fifty-seven-year-old man told his wife one Saturday night nearly a year ago. While she'd been at the theater—they'd bought the tickets weeks earlier—he'd had to crawl up the stairs on his hands and knees to get back to bed. Terrible bone-shaking chills tore through him, despite the thick layer of blankets. The shivering was followed by sudden blasts of internal heat and drenching sweats that made him kick off the covers, only to have to haul them back up as the cycle repeated itself.

You really need to go back to the ER, his wife told him. The frustration and worry were clear in her voice. He'd already been to the emergency room three times. They'd given him some intravenous fluids and sent him home with the diagnosis of a viral syndrome. He would start to feel better soon, he was told each time. But he hadn't.

This all began nine days before. That first day he called in sick to his job as a physical therapist. He felt feverish, as though he might have the flu. He would drink plenty of fluids and take it easy and go back to work the next day. But the next day he felt even worse. That's when the fevers and chills really kicked in. He alternated between acetaminophen and ibuprofen, but the fever never let up. He started sleeping in the guest room because the sweats soaked the sheets and the chills shook the bed, waking his wife.

After four days of this he made his first visit to the Yale New Haven Hospital emergency room. He was already being treated for a different infection. Three weeks earlier he'd developed a red swollen elbow and gone to an urgent care center, where he was started on one antibiotic. He took it for ten days, but his elbow was still killing him. He went back to urgent care, where he was started on a broader-spectrum drug, which he was nearly done with. Now his elbow was fine. It was the rest of his body that ached.

But his flu swab was negative. So was his chest X-ray. It was probably just a virus, he was told. The antibiotics he was already taking would kill just about any of the likely bacteria. He should just take it easy till it passed. And come back if he got any worse.

The next day his fever spiked to 106. And so he went back to the ER. There he found a mob scene—crowded with people who, like him, felt like they were sick with the flu. It would be hours before he could be seen, he was told. Discouraged, he went home to bed. A nurse from the ER called the next morning. Could he come back now that the ER was more manageable? He was happy to return.

He may not have the flu, he thought, but he was sure he had something. But the ER doctor couldn't find it. He didn't have any chest pain or shortness of breath. No cough, no headache, no rash, no abdominal pain, no urinary symptoms.

His heart was beating hard and fast, but otherwise his exam was fine. His white-blood-cell count was low—which was a little strange. White blood cells are expected to increase in the setting of an acute infection. Still, a virus can cause white counts to drop. His platelets—the tiny blood fragments that form clots—were also low. That can also be seen in viral infections, but it was less common.

The ER staff sent the abnormal blood results to the patient's primary care provider and told the patient to follow up with him. He'd been trying to get in to see his PCP, but the doctor's schedule was full. It was the worst flu season in years. When he called again, he was told that the soonest he could be seen was the following week.

The office agreed to order blood tests to look for Lyme and other tick-borne infections. This was Connecticut, after all. He dragged himself to the lab and waited for his doctor to call with the results. The call never came. In his mind, he fired his doctor. He'd been sick for over a week and they couldn't see him, couldn't even call him with the lab results he'd asked for.

He again went to the ER on Sunday, the morning after his wife returned from the theater and insisted he go back. His previous visits and lab abnormalities caught the attention of the physician's assistant on duty that morning. She ordered a bunch of blood tests—looking for everything from HIV to mono. She ordered another chest X-ray and started him on broad-spectrum antibiotics, as well as doxycycline, an antibiotic for tick-borne infections. He was given Tylenol for his fever and admitted to the hospital. As he prepared to leave the ER, the flu test came back positive. He was pretty sure he didn't have it; he'd never heard of a flu lasting this long. But if he could stay in the hospital, where someone could monitor if he got worse, he was happy to take Tamiflu.

The lab called again later that day to say that the test had been read incorrectly; he did not have the flu. But by then other results started to come in. It definitely wasn't his elbow—according to the patient, the orthopedic surgeon who saw him in the ER, and an X-ray. He didn't have HIV; he didn't have mono, or Lyme; he didn't have any of the other respiratory viruses that, along with influenza, had filled up much of the hospital. Yet, after a couple of days, the patient began to feel better. His fever came down. The shaking chills disappeared. His white count and platelets edged up. It was clear he was recovering, but from what? More blood tests were ordered, and an infectious disease specialist consulted.

Gabriel Vilchez, the ID specialist-in-training, reviewed the chart and examined the patient. He agreed that it was most likely that the patient had a tick-borne infection. The hospital had sent off blood to test for all the usual suspects in the Northeast: Lyme, babesiosis, ehrlichiosis, and anaplasmosis. Except for the Lyme test, which was negative, none of the other results had come back yet. Vilchez thought that given the patient's symptoms—and his response to the antibiotic—it would turn out that he had one of them.

And yet, all the results for tick-borne infections were negative. But there were other tick-borne diseases, less common in the Northeast but still possible. To Vilchez, the most likely was Rocky Mountain spotted fever (RMSF)—though it's much more common in the Smoky Mountains than the Rocky Mountains. The spotted fever part, the rash, was seen in most but not all cases. It's unusual to find the infection in Connecticut, but not unheard of. Vilchez sent off blood to be tested for RMSF and to retest for the other infections. The following day the patient felt well enough to go home. A couple of days later he got a call. He had Rocky Mountain spotted fever.

The patient, it turned out, had the misfortune of experiencing fever and flu-like symptoms in the midst of a flu epidemic. Under these circumstances the question quickly becomes not What does he have? but Does he have the flu? Once you get to no, it's hard to go back to the broader question.

For the patient, recovery has been tough. Though the doxycycline helped with the acute symptoms, it took months before he could resume his usual patient load at work. He just didn't have the strength or the stamina to get the job done. He feels that the illness brought him as close to dying as he'd ever been. Indeed, RMSF is the most dangerous of all the tick-borne infections, with a mortality rate as high as 5 percent even with current antibiotics.

One thing he was certain about, however. He needed a new primary care doctor. And he got one.

The Flu That Stayed

Dr. John Henning Schumann was worried. His best friend from college, something of a hypochondriac, often called him with medical questions. A few weeks earlier he'd mentioned a viral illness—the basic fever and malaise, nothing too concerning. But now he heard from a mutual friend that those symptoms never went away—and that was concerning. Schumann told his friend to see his doctor immediately.

A few days later Schumann received an email from his feverish friend. He was in the hospital. He had seen his doctor, who sent him for a CT scan of his abdomen. The scan revealed a softball-size mass in his liver. His doctor sent him to Mount Auburn Hospital in Cambridge, Massachusetts, for further testing.

Dr. Andrew Modest was the internist assigned to his care. Before going in to see the new patient, he scrolled through his electronic medical record: He was forty years old, a university professor, and completely healthy—until now. The blood tests revealed a mild anemia, and, of course, the CT scan showed the mass.

The patient sat comfortably in his hospital bed with a computer on his lap. He looked a little pale but otherwise appeared well. "I'm writing about my illness for my friends and family," the patient announced cheerfully. "Is that okay with you?"

The fevers started a week after he returned from a conference in Switzerland, the patient told Modest. They only came at night, but they came every night. First the fever and then hours later he would break out in a heavy sweat. Sometimes it was so bad that he had to change his pajamas and sheets. He had a tickly kind of cough, too. Other than that, he had no complaints. He had lost fifteen pounds over the past month but assumed that was because he had started a new diet.

On exam, the doctor found nothing out of the ordinary. The patient had no fever, although his temperature had gone up to nearly 102 the night before. His heart rate and blood pressure were normal. So was everything else.

The patient had had fever for four weeks and lost a good deal of weight, which the doctor thought was unlikely to be because of the minor dietary changes the patient reported. Was this an infection? Maybe, although he didn't look sick. Was it some sort of autoimmune disease like lupus? Or was it some sort of cancer? Any of these were possible.

He also had the huge mass in his liver; was that the source of the fever? Or was the mass what is known as an incidentaloma—an abnormality found when searching for something else? The size alone suggested it had been there for quite some time—probably years. Why would he get a fever now? And if it wasn't the liver, then what? Tick-borne diseases like Lyme and anaplasmosis could cause this kind of persistent nocturnal fever. So could HIV, TB, hepatitis, and dozens of other infections.

Modest sought out a radiologist to review the images of the liver mass with him. The radiologist's first thought was that this could be a very big hemangioma—an abnormal but benign collection of blood vessels. But these tumors usually have smooth borders, the radiologist added, and this one didn't. Besides, hemangiomas typically don't cause fevers.

What else could it be? the doctor asked. There is a malignant form of hemangioma, the radiologist added thoughtfully. Known as an angiosarcoma, they can cause fevers but are extremely rare. Or it could be a benign hemangioma with some kind of infection within it. That could cause a fever and would require immediate antibiotics. But in order to make this diagnosis, they needed to see what kind of fluid was in this mass. If there was pus, they would have to drain it and start antibiotics. If no infection was seen, they could hold off on the antibiotics, at least for the time being.

That afternoon the radiologist inserted a long needle into the mass in the patient's abdomen. Once the needle was in place, the doctor pulled back on the syringe and the chamber filled with dark red blood. The lab quickly reported that there was no pus in the blood and no evidence of an infection. Modest shared the news with the patient. He told him that he wasn't sure what the source of his fever was. An infectious disease doctor would examine him later that day and a gastroenterologist would come by over the weekend. Modest would be back Monday.

That night the patient was so worried he could barely sleep. The next morning he called Schumann. He had had many tests, he said, including CTs and MRIs; he had been poked, prodded, and stuck. As a result, there were many things his doctors were pretty sure he didn't have—it didn't seem to be cancer; they still couldn't find an infection; it wasn't HIV or hepatitis or lupus—but none of his doctors had been able to figure out what he did have. And that uncertainty was scaring him.

Schumann was worried, too. He lived nearly a thousand miles away—too far to come and see his friend. Besides, the patient was posting all his test results for his friends and family, so Schumann was following the case from a distance, and he still had no idea what was going on. If he was the second opinion, then the patient clearly needed a third. Suddenly Schumann had an idea. What if they opened the mystery of his symptoms to other doctors? What if they put his case on the Internet—on a blog read mostly by doctors—and let some new eyes and brains work on this problem? The patient was excited by the idea.

That afternoon Schumann put the case on his blog (www.glasshospital.com) and contacted Kevin Pho, who has a popular medical blog (www.kevinmd.com), who then posted it as well. Within hours, a dozen comments were posted in response. Several pointed to a series of reports similar to this very case: patients with large hemangiomas and persistent nightly fevers. In several cases, simply removing the mass stopped the fevers.

Hemangiomas are the most common benign tumors found in the liver. Most remain small and asymptomatic. But occasionally they can become quite large, and when that happens, patients often complain of pain or a sense of fullness. In very rare cases, and for reasons that are not well understood, these tumors can cause fever, weight loss, and anemia—the same symptoms that this patient had.

As soon as Schumann saw these case reports, he was hopeful that this was the right diagnosis. The patient was optimistic, too.

Modest never saw those case reports. Nevertheless, he arrived at the same diagnosis, albeit through a more traditional route. The patient was seen by Dr. Frederick Ruymann, a gastroenterologist. Ruymann had seen a case just like this years earlier and recognized it immediately. Still, because there was no way to know if this benign tumor was the cause of the fever short of taking it out, Modest had to be sure that he had ruled out as many alternative diagnoses as he could before turning his patient over to the surgeons. All his tests were unrevealing, and finally, by midweek, Modest was comfortable that the symptoms were probably caused by the hemangioma.

The patient had the hemangioma removed in April. Although the recovery from the operation was more rigorous than he expected, he is now back to his old self. No more fevers, no more fatigue; even the cough disappeared.

In medicine, doctors accept that no one knows everything. Our knowledge is shaped by experience, training, and personal interest. We all reach out to our community of doctors when we are stumped. Usually it's to our friends and colleagues, but the Internet offers the possibility of a broader community—a sea of strangers linked by our medical curiosity and by our keyboards.

Burning Up at Night

Her mother had fallen and was too weak to get up, the kind voice on the phone explained. The call from the young woman's aunt sent her hurrying to her mother's home, just a town away in rural Alabama. The aunt had found her sixty-eight-year-old sister naked and confused in the living room. She'd called her sister, as she had every day since her sister had gotten so sick, and when she didn't answer, she was worried and drove over to the house. Her sister was still confused but fully dressed by the time her niece arrived.

Although her mother had been sick for several years, the daughter was still shocked to see the pale, wasted shadow she had become. She'd been to the ER of the local hospital many times. She had even seen specialists in Tuscaloosa. But no one seemed to have a clue what was wrong with her.

As the EMTs loaded her mother into the ambulance, the daughter asked if they could drive all the way to Birmingham. When she was pregnant with her triplets the year before, she'd traveled the fifty miles to see specialists at the University of Alabama Hospital in Birmingham. Maybe the doctors there could help her mom.

The doctors in the emergency room in Birmingham gave her mother some fluids to bring up her blood pressure, and she perked up a little. More important, they persuaded the older woman and her daughter to follow up at the university's outpatient clinic.

Dr. Jori May, an internist in her second year of training, introduced herself to the thin, pale woman and her two daughters a month later. They offered May a thick stack of medical records they'd brought, and she put them aside to look at later. First, she needed to understand what had happened.

It started years earlier, the older woman told her. Almost every night, she would get these crazy fevers. First came bone-rattling, shaking chills; she couldn't get warm even under a pile of quilts. Then suddenly she would be roasting hot, with sweat pouring off her. Her temperature would spike to 102 or 103. And her whole body would hurt, right down to her bones. She popped Tylenol constantly for the fever and the pain.

Then an hour after her temperature spiked, she would start to feel sick and throw up until she had nothing left in her. This happened almost every night.

During the day, she felt weak and tired, and her bones hurt. It made any movement painful. Her doctors called it fibromyalgia. She also had a rash. Hives, the doctors told her. It didn't itch, and no one could figure out why she had it. And, her daughters added, she had no appetite. The very thought of food made her want to vomit, the older woman told May. She'd lost over eighty pounds this past year.

May could see how the patient's clothes, her eyes, and

even her skin looked a couple of sizes too large. Otherwise her exam was uninformative. She didn't have a fever, and she didn't have a rash. May told the patient she would go through the stack of records and come up with a plan.

Reviewing them, May saw that the patient had a persistently elevated white-blood-cell count. Normal is under ten; the patient's was at nearly twenty—and had been for a couple of years. CT scans showed enlarged lymph nodes throughout her body. These findings could be from a chronic infection. Or from a cancer. But her hometown doctors found neither.

May decided to think about illnesses that the patient's first doctors hadn't tested for. The woman needed to be checked for HIV; individuals over fifty-five are thought to make up a quarter of all cases—if you counted the undiagnosed as well as those with that formal diagnosis—and these older patients are much less likely to be tested for it. Another possibility was syphilis, called the great imitator for its variable presentations. And given her persistent gastrointestinal problems, May would look for celiac disease. She also sent off a test to look for a type of blood cancer called multiple myeloma, which attacks the blood and bones and is seen in patients over fifty.

May waited anxiously as the results came back. It wasn't HIV. It wasn't syphilis or celiac disease. The patient didn't have multiple myeloma, either, though that test, which measures levels of one part of the immune system known as antibodies, was abnormal; one type of antibody, known as IgM, was high. May referred the patient to an infectious disease specialist, who found no infection. The oncologist found no cancer. And the dermatologist merely confirmed what May already knew—the patient had hives, and it wasn't clear why. She presented her puzzling patient to every smart doctor she knew when she walked down the hospital hallway and at educational conferences. Yet after seven months of testing and referring and discussing, May was no closer to a diagnosis than she was on day one.

It was part of May's weekly routine to check the patient's chart for any new consultant notes or test results. One afternoon she was surprised to see an eleven-page note from a pathology resident named Forest Huls who, as far as she knew, was not even involved in the case. It was a meticulous summary of all the patient's symptoms as well as the many tests performed so far. Huls went on to suggest that her patient had a disease May had never heard of—Schnitzler syndrome. It was, as the resident described it, a rare and poorly understood immune disorder.

According to current thinking, in Schnitzler syndrome the most primitive part of the immune system—a type of white blood cell known as the macrophage—goes wild and instructs the body to act as if it is infected. The body responds with fever and chills, a loss of appetite, flu-like body aches, hives, and high levels of one specific type of antibody: IgM. Exactly why and how this occurs is still unclear.

The disorder was first described in 1972 by the French dermatologist Liliane Schnitzler, who had identified five patients with hives, episodes of prolonged fever, bone pain, and enlarged lymph nodes. These symptoms, plus an elevated level of IgM, Schnitzler proposed, defined a new disease.

May didn't know Huls personally, but she'd heard of him. Although still in training, he had a reputation for finding cases that stumped others and figuring out the diagnosis. "When I see people suffering and I know that if I took the time and effort, I could figure it out," he told me, "then I have to do something." He looks for unexplained pathological findings—in this case, the high level of IgM.

Huls hadn't heard of Schnitzler syndrome, either. He came upon it by using the database PubMed to look for a disease that matched the patient's symptoms. He made a list of her symptoms and abnormalities. To get the full picture, he combed through her earlier electronic medical records, now archived in an old electronic warehouse, and found that her symptoms had started maybe a decade earlier. Then he looked for a disease that fit. It took hours before articles on this strange disorder began to appear. As he read, he suspected that she had it.

After reading Huls's note, May looked up Schnitzler syndrome. The descriptions she read of patients with the disorder matched this woman exactly.

It was an important diagnosis to make, in part because there is now a very effective treatment. The disease causes the macrophages to overproduce a protein called interleukin 1.

This is the protein that tells the body to act sick—develop a fever, body aches, and all of the rest of the flu-like symptoms so characteristic of Schnitzler's. And a few years ago, a drug company developed a medication that specifically inhibited the actions of this particular protein. When the woman's insurance company refused to pay for this new and very expensive drug, May appealed to the manufacturer, and they agreed, after several months, to provide it. Once she started taking the medicine, the shaking chills and fever disappeared. So did the nausea and vomiting, the hives, and the bone pain.

Looking back at her life with this illness, the patient can barely recognize herself. Before her illness, she'd prided herself on her get-up-and-go and her disinclination to sit still. All those years stuck on the sofa and ultimately in bed, too sick, too weak, and in too much pain to move, seemed like a chapter in someone else's life.

As for Huls, he finished his fellowship and has returned to the university of Alabama. In his new job he will take on the toughest diagnostic cases—cases that will certainly challenge his curiosity and his skills as he works to solve them.

Sick at the Wedding

"Either you are getting in the car with me to go back to the hospital, or I'm calling an ambulance," the woman announced to her thirty-eight-year-old husband. He'd been home from the hospital for only a day, but he looked sicker than ever. Though she couldn't bring herself to say it, she was worried that he might be dying. And so was he.

It started at his younger brother's wedding about a week before—a destination event in Colorado. Almost from the moment he stepped off the plane, he felt awful. His head throbbed. His body ached. His eyes felt puffy, and his whole face looked swollen. When he went to bed that first night, he tossed and turned, sleepless. In the morning, when he pulled himself out of bed, the sheets were soaked with sweat.

At first he wrote it off to altitude sickness. The resort was in the mountains, high above sea level, and he had never been up at that altitude. Though his wife and two children felt fine, others in the wedding party were feeling the effects of the elevation. One bridesmaid fainted at the reception; an elderly aunt from Texas had to leave early.

The afternoon wedding service seemed to last forever. His tuxedo felt like a straitjacket; his chest felt tight, and he could barely breathe. By the time the reception dinner started, he felt terrible. He shook with violent chills, and his head was pounding. His neck hurt, and he could hardly swallow. His wife asked the host to change the order of the toasts so that he could give his early. Then he went to the hotel and climbed into bed.

He figured he would feel better when they got down to Denver. But he didn't. And back in Boston, at sea level, he still felt awful. His wife reluctantly left him in the city, where he had a flight to catch early the next day, and drove to their home, an hour away.

When he was alone in his hotel room, his symptoms seemed even scarier, and late that night he took a taxi to Massachusetts General Hospital. Because of his chest tightness, he had an EKG. To his surprise, it was abnormal, and he was rushed to the cardiac-care unit. The doctors were sure he hadn't had a heart attack, but something had damaged his heart. After dozens of tests, they told him he had myocarditis, an inflamed heart muscle, but they weren't sure why. They searched for a cause. Myocarditis often stems from a viral infection. But bacteria can also infect and injure the heart; they looked for strep and other possible culprits but didn't find them. They were worried that he had picked up a tick-borne infection in rural Colorado. None of the tests were positive, but after four days they sent him home to finish up a week of an antibiotic called doxycycline, just in case.

At home, he went to bed, hoping he was on the mend. His wife wasn't so sure. The next day, when she looked in on him, she was frightened by how very sick he looked. He was pale and sweaty, the way he'd been in the mountains. The shaking and fever were back. His headache was terrible; the day before it was so bad that he'd cried with pain—something she'd never seen before. The prospect of an hourlong drive back to Mass General seemed daunting. That's when she told him he had to go to the hospital and decided to drive him to Anna Jaques Hospital, a community hospital one town over in Newburyport.

It was late by the time they arrived at Anna Jaques, and the emergency room was quiet. Dr. Domenic Martinello knocked at the entrance to the cubicle where the patient waited. His wife looked up expectantly, her face tight with exhaustion. The patient lay motionless on a stretcher; his eyes were sunken, and his skin hung off his face as if he hadn't eaten much recently. His voice was soft but raspy, and every time he swallowed, his lips tightened in a grimace of pain. Together husband and wife recounted the events of the past few days: the wedding, the fevers, headaches, pain in his neck and throat, the four days in the hospital in Boston.

It was certainly a confusing picture, and Martinello wasn't sure what to make of the diagnosis of myocarditis. In any case, the man had no chest pain now. Only the headache, the sore neck, and the painful throat. The doctor quickly examined him. The patient's skin was warm and sweaty, and his neck was stiff and tender, especially on the right. Martinello was going to approach this systematically, he told the couple. First he would get a head CT scan, then a scan of the neck. Then he would do a lumbar puncture—a spinal tap. He felt optimistic that one of those tests would provide an answer.

The head scan was normal. There was no tumor, no

blood clot, and no sign of increased pressure. Because of the patient's tender neck, Martinello wondered whether he had an abscess there. It was the right question, though the result was not what Martinello expected. There was a small abscess. More worrisome, there was a blood clot in the patient's internal jugular vein on the right side. It was a sign of Lemierre's syndrome, a rare infection that Martinello had seen only once before.

Lemierre's was best described in the 1930s by André Lemierre, a French researcher who reported twenty cases of this previously undiagnosed condition. The patients started off with a sore throat and subsequently developed a clot in the jugular vein. The clot would frequently break apart, and the pieces—each containing some of the bacteria that caused the infection—would carry the bug to other parts of the body, mostly the lungs but occasionally bone, brain, or other organs. Typically, this infection is caused by an unusual bacterium called *Fusobacterium necrophorum.* No matter which bug caused it, Lemierre's was almost a death sentence in the pre-antibiotics era. Even now, up to 18 percent of patients with Lemierre's will die.

In this case, blood cultures suggested that the infection had started with a disease that is far more common and much less feared: strep throat. There are millions of cases of streptococcal infection in this country every year, usually in the throat or on the skin. A tiny fraction of the time, these bugs can invade the surrounding tissues and cause a life-threatening illness, as they had in this patient. Both the Lemierre's and the myocarditis were caused by this strep throat gone wild. This kind of invasive infection must be treated with antibiotics, but the antibiotic the patient was taking for a possible tick bite, doxycycline, isn't effective against *Streptococcus pyogenes*, the species of streptococcus that causes strep throat. He had been tested for this organism at Mass General, but the result was negative. It's not clear why, but no test is 100 percent accurate. By the time the patient came to Anna Jaques Hospital, the bacteria was in his blood and so easily found.

Now that Martinello knew what was making this man so sick, he was worried that his small community hospital was not prepared to care for him. They didn't have the kind of specialists he needed on call. Martinello arranged for the patient to be transferred to a sister hospital, the Beth Israel Deaconess hospital in Boston. At Beth Israel, the patient was closely monitored by infectious disease specialists and ear, nose, and throat surgeons. He continued on antibiotics for six weeks and started a course of a blood thinner to keep the clot from growing or spreading.

It took months, but the patient completely recovered. Looking back, he remembered that his throat was painful, but, he told me, it seemed insignificant compared with the shaking chills, fever, and headache. "I thought of it as kind of a sidebar, when in fact it was the main event," he said. After their scare, he and his wife read up on the illness. Now their family motto is: Take strep seriously.

Forgotten Triggers

The patient was an attractive woman in her late seventies, her face deeply creased by years and tobacco. Her hair was white, her eyes a pale blue. But it was her skin color that you noticed first: Her face and arms were a deep bright red, as if she were sunburned, though it was February in Connecticut. She looked at the intern standing in front of her. "Back already?" the patient growled.

The intern stepped forward. She was a woman in her late twenties, confident and matter-of-fact. "Yes, ma'am," the intern said. "I told you I would be back later with the team, and here we are." When prompted, the patient repeated her story. She'd been fine until a couple of days ago, when she started feeling "really rotten," weak, and achy. "Then I got this rash and the chills," she said, and she noticed that she had stopped going to the bathroom. She called her doctor and then her son, who brought her to the hospital.

In the ER, she had a temperature of 102.8. Her blood

pressure was quite low—in the eighties—and her heart was racing. Lying on the stretcher, she looked tired. Her lips were dry, and oddly, when she stuck out her tongue, it trembled as if the effort were too much for her. Her lungs were quiet and clear. Her abdomen was soft and not painful.

The rash glowed a uniform red on her face and arms, but as we examined her we noticed that on her trunk and back it looked a bit different, consisting of a multitude of tiny, raised bumps, each surrounded by a small circle of redness. The only areas of her body completely spared by the rash were the palms of her hands and the soles of her feet. "Oh, my God, it itches," she groaned as she caught herself scratching.

Although the patient described herself as "pretty healthy," she had quite a few medical problems. After a lifetime of smoking cigarettes (she'd quit four years before), she had serious lung disease. She also had a history of coronary-artery disease. And she had just been in the hospital the previous month with pneumonia.

She recited a long list of the medicines she took: beta blockers, aspirin, and nitroglycerin for her heart, inhalers for her lungs, but none were new, and she'd been taking all of them without any difficulty.

Patients as sick as this one usually have blood drawn and sent to the lab even before a doctor sees them. So by the time the team saw her, we knew she had a high white-blood-cell count and there was evidence that her kidneys were not working—not working at all.

There is a principle in medicine, borrowed from philosophy, that when possible, you should strive to come up with the simplest possible explanation for the phenomena you observe. In medicine, that means we try to find a single diagnosis to explain all that we see in a patient.

Occam's razor, it's called—the art of shaving the diagnosis down to the simplest, most elegant solution. It is one of the great pleasures of medicine, but in this case, it wasn't going to be easy.

Here's why: The patient had a fever, low blood pressure, and an elevated white-blood-cell count. That combination meant infection until proved otherwise. But if this were an infection, how did the rash fit in? And what about her failing kidneys?

There are some serious and unusual infections that result in both a fever and a rash. Toxic shock syndrome is one; Rocky Mountain spotted fever is another. But although this patient was clearly sick, she didn't appear quite as ill as she'd be if she were suffering from one of those diseases, whose rapidly progressive courses are among their most deadly characteristics. Her vital signs, although abnormal, had been stable since her arrival. Moreover, our patient's rash was clearly itchy; not so the rash associated with those infections.

A severe infection that caused very low blood pressure could make your kidneys stop working, simply because they weren't getting enough blood. But our patient, although noticeably uncomfortable and tired, was thinking clearly, and that suggested that while her blood pressure was low, it was high enough to get blood to the important organs. If she could think, she should be able to make urine, and yet she wasn't.

We gave her fluid intravenously, but even after a couple of hours, no urine appeared. However, giving her fluids did bring her blood pressure back to normal, and that was an important change. Fever and hypotension—most of the time that combination means infection. Fever alone, with a normal blood pressure, might be caused by infection but could also have other causes.

So now her symptoms were adding up in a slightly different way. She had this very high fever; she had an itchy rash; and she had very impressive kidney failure. If this weren't an infection, what could it be?

Certainly the most common noninfectious cause of fever, at least in the hospital, is medications. Many drugs, especially antibiotics, can cause a type of allergic reaction that includes fever and often a rash. But our patient said she wasn't on any new drugs, and none of her old drugs seemed likely candidates. There are certain types of severe arthritis that can cause a fever; some even have a rash, and a few can on occasion cause kidney problems. But her physical exam showed no evidence of this. Cancer, most commonly lymphomas, can also cause fever, but at this point she had nothing suggestive of malignancy.

The team returned to the patient's bedside. It was late by then. A tired-looking man wearing a somewhat wrinkled suit sat by her bedside. He introduced himself as her son, and the young resident immediately asked if his mother had started any new medicines. "No, not recently," he confirmed. The resident explained to the patient and her son that we didn't know exactly what had caused her to become so ill. "She probably has an infection," the resident said. "But it's not clear yet why her kidneys aren't working."

As we prepared to leave, the son spoke again. "Last month my mother started a medicine for her gout," he offered, "but that's not really new, is it?"

The medicine was allopurinol, a very effective medicine to prevent gout attacks, but also one well known to cause allergic reactions. From the son's perspective, it didn't seem a new medicine. And from our patient's perspective, it was so new, she forgot she was taking it. From our perspective, the timing was just right, completely typical of this kind of drug reaction, known as allergic interstitial nephritis. This unusual and complex allergic reaction—with its classic triad of fever, rash, and renal failure—was a perfect explanation. She would need a biopsy of her kidney to confirm this diagnosis, but it seemed the likely culprit. The fit was just too good.

This is one of the most gratifying moments in medicine. You have a patient with complicated symptoms that could fit together in lots of ways, and it seems there is no simple and elegant answer. Diagnoses, like people and their lives, are often complicated and messy. But every now and then, there is the patient whose history, signs, and symptoms suddenly come together perfectly. You get one piece of information and suddenly you see the pattern and know the disease. The patient gets a perfect diagnosis; the doctor gets to take pleasure in finding it.

Our patient was started on dialysis the next day, because her kidneys were too damaged to work. The biopsy of her kidneys confirmed our diagnosis, so she would probably need dialysis for only a few weeks. Once the offending medicine was stopped, she rapidly improved.

By the end of the week, she looked like a new woman. Her fever was gone. Her bright red skin faded, and all that was left of her rash were a few faded scratches. When she started to complain about the food, we knew she was probably ready to go home.

A Killer Flu

The wintry late-day sun flooded through the windows at South County Hospital in Wakefield, Rhode Island, as the middle-aged man and his wife entered the room. The man's mother, a tiny ninety-three-year-old woman, sat slumped amid a chaos of bedcovers. They had come from St. Louis after the youngest of the man's siblings called to let him know that their mother was deathly ill. Seeing her now, a pale, silent version of her energetic self, he feared he would have to wear the suit he'd brought after all.

She'd been in this hospital for almost a week, though the symptoms started a week before that. It was that Saturday morning when she noticed that she felt a little tired. By midday she was cold and flu-y. Her body ached all over—especially her back. And she had a fever. A neighbor took her to the ER. For reasons she couldn't remember, they ended up at a hospital a couple of towns over. There, blood was drawn and a CT scan was done to look for the cause of her symptoms. Nothing was found, so she was sent home with a medication for her painful back.

On Tuesday, she went to her primary care doctor. He looked her over and reviewed the ER records. He wasn't sure what more he could do. Another son, who lived nearby, and his wife started sleeping at his mother's house. They were worried about her. That this fiercely independent woman who lived alone, who still cut her own wood for the stove and drove everywhere, felt sick enough to go to the ER indicated to them that despite what all the doctors might say, she was seriously ill.

When she didn't get better after a few days, they carried her to the car and went to South County, the hospital they knew best.

The doctors at South County weren't sure what was going on, either. The patient felt sick: She was tired, her back was killing her, and she felt weak all over. She looked sick, too: pale and frail. On exam, her temperature was up and her blood pressure was down. She had a faint rash on much of her body, and—this scared her son the most—she was confused. But her white-blood-cell count—an indicator of infection—wasn't elevated, and the doctors couldn't see any obvious source of infection. Blood tests showed no signs of the most common tick-borne diseases. A chest X-ray was normal. An ultrasound of her abdomen was, too. Blood was also analyzed to see if any bacteria would grow, and she was admitted to the hospital under the care of Caroline Jenckes, an experienced nurse practitioner.

Jenckes spent the next few days looking for an infection she was certain was there. She ordered an MRI of the spine in search of an abscess, but found nothing. The patient's gallbladder was carefully evaluated; inflammation there can trigger a fever along with pain that travels to the back. A couple of days in, a CT scan of the patient's chest suggested the possibility of pneumonia. Dr. Fred Silverblatt, the infectious disease doctor Jenckes consulted, didn't think the subtle findings could be the cause of her symptoms. And the patient's fever was already coming down. Still, he agreed to start her on broad-spectrum antibiotics.

Finally the medical team saw what seemed significant evidence that the patient was getting better. Her fever came down, and her blood pressure returned to normal. Her back pain was subsiding. Despite all that, the son knew that his mother was not really recovering. He and his wife had been taking turns staying with her, day and night, and she was nothing like her old self; she still seemed quite sick. Despite the antibiotics, she remained tired, weak, and practically speechless. The son called his sister in Maryland and older brother in Missouri to tell them that this might be the end, and the siblings hurried back to their hometown.

The frail woman didn't open her eyes when her older son entered the room. He leaned close to give her a kiss and straighten her up in the bed. Did she have flying-squirrel fever, he asked lightly, referring to their effort the autumn before to rid her attic of the pesky rodents. Back then she'd told him how cute the big-eared babies were, creatures taken out of the attic by the exterminator on their way to relocation. He thought she smiled a bit at his joke, just a whisper of her usual good humor. But as he made the joke, he suddenly had a thought: Could those squirrels have anything to do with this strange illness that no one could figure out?

It was an odd thought but the kind of connection that inspired the son, so he found a computer in the hospital and searched online a bit. The first pages carried mostly ads for services to get rid of the pests. But then he found something: a short article from the Centers for Disease Control and Prevention linking flying squirrels with something called epidemic typhus. Further reading revealed that the symptoms of typhus—fever, body aches, rash, and confusion—resembled those his mother had. But typhus was a rare infection. There had been fewer than a hundred cases reported to the CDC over the past forty years. Still, the son printed out the article and went in search of Caroline Jenckes. He explained that his mother's house had been infested with these animals. Jenckes was intrigued with the idea. Certainly all the studies they'd done to find the source of the woman's symptoms had been unrevealing. She took the article to Silverblatt. It seemed to him a perfect fit: the symptoms, the exposure, the minimal response to the antibiotics. He read up a little more on the infection before starting the patient on the appropriate antibiotic—doxycycline—and sending her blood to the CDC for confirmation.

Epidemic typhus is an old disease. Since the Middle Ages, periodic outbreaks of this infection have killed millions of people. Just after World War I, an outbreak in Russia killed three million. The infection is often transmitted through contact with body lice. Modern sanitation has significantly reduced the incidence of both the infection and the carrier. In the United States, most cases of typhus have come from exposure to flying squirrels. It's not clear how the bacteria get from the rodents who harbor the bug to the humans they infect, but the louse once again plays a role. Squirrel lice do not tend to bite humans; instead, it is thought that exposure occurs when bacteria in lice excrement are inhaled.

After twenty-four hours on the new antibiotic, the

woman was transferred to a rehabilitation facility. Within days she began to act more like herself. She was irritated that she had been placed on the floor designated for the sick and dying. She was certain, she explained to anyone who would listen, that this was not where she belonged. She swears she didn't really start to get better until she moved to the floor for healthier patients.

After a few weeks she was well enough to go home. It took almost that long for the test results to come back confirming typhus. While she recovered, her children arranged for an exterminator to get rid of any squirrels that might have returned and to seal up any possible ports of reentry.

Families are an essential source of information about patients and the world they live in. They don't usually make the diagnosis, but they can provide answers to questions doctors hadn't considered asking. In this case, the knowledge of this woman's exposure and its associated infection may have saved her life. Untreated epidemic typhus can kill up to 30 percent of patients, and the very old are at the greatest risk.

Years later, if you were to ask this woman how she's doing, she'd immediately inform you that she's just great—because the more she's on her own, the better she feels.

PART II

A Pain in My Belly

Excruciating Episodes

The stomach pain would come with unbearable intensity, last a day or two, and then disappear. "I just can't go on this way." The patient, a tall, gaunt teenager with a short, dark crew cut and a worried look on his face, spoke quietly but with a sense of urgency. "My stomach—it hurts me so much. I can't go back to college until I know what's happening." His mother, young and slim and clearly anxious, nodded. "It's been going on too long," she added. Dr. Kiran Sachdev, a gastroenterologist, agreed. It had been going on far too long.

The patient first came to see her three months earlier. He described then these attacks of intense abdominal pain that he got every couple of months or so. The pain was sharp, he said, unrelenting. During these attacks he couldn't eat, he couldn't walk, he could barely stand. Then, after a couple of days, he would get better. He didn't know what brought on the pain or why it went away. But it had been going on for almost a decade, and he just wanted it to stop.

It all seemed to start after his appendix ruptured when he was eleven, his mother told Sachdev. He needed two operations and was in the hospital for almost three weeks. A month or so later, her son had his first attack of this mysterious pain. Initially, they went to the ER with every severe episode. The doctors there never figured out what was wrong, so eventually they just dealt with it at home. A few years later, during a particularly bad episode, his mother took him once more to the emergency room. A surgeon there told her that the scars from his appendicitis could be causing the bouts of pain by intermittently blocking his digestive tract, a relatively common complication of abdominal surgery. He recommended another operation to remove the scar tissue. Less than a month after the patient had that operation, the pain came back—just as bad as ever. The surgeon was baffled. "He told me I wasn't eating enough fiber, but I knew that wasn't right," the young man said. He tried Metamucil and increased the fiber in his diet, but the attacks kept coming. Despite these recurrent episodes, he graduated from high school and started college.

Away at school, he continued to have attacks. During his first two years there, he told Sachdev, he probably spent more time sick in bed than actually sitting in class. But he was determined not to let the attacks limit him.

The patient had no other medical history except for unidentified food allergies that caused his feet or hands to swell occasionally. For that he just took an antihistamine. He didn't smoke, drink, or use illegal drugs. On physical exam, Sachdev found the patient to be thin but healthy. His abdomen was flat, with good muscle tone and without pain or any palpable abnormalities. All the routine blood work had been normal, too. A CT scan, done at the time of his most recent hospital stay, did show an abnormality: fluid outside the organs in the abdomen. The ER doctors hadn't known what to make of that. Neither had Sachdev, but it did suggest that whatever the cause of the pain, it was in his gut and not in his head.

When she first saw him, Sachdev thought that he might have irritable bowel syndrome, a disorder in which the gut overreacts to normal stimuli like food or gas or stress. It usually causes an intermittent, crampy pain. She put him on a medicine that prevents spasm, and he had done pretty well, until last week, when he ended up in the hospital once more. Again a CT scan showed free fluid in the abdomen. And again his symptoms resolved quickly. He was discharged within forty-eight hours and was now back to see Sachdev the following week. It wasn't irritable bowel. So what was it?

When doctors talk about the art or science of diagnosis, they often break it down into two distinct processes: One depends on pattern identification—you see a patient, recognize the signs and symptoms of some known disease, and make your diagnosis. Maybe you get a test to confirm what you already know. Maybe you don't. In either case, you are satisfied that you have figured it out.

Then there are the rest of the patients—the ones with symptoms that don't match any of the patterns you know. In these cases, many doctors I have spoken to describe developing their own hierarchy of possible diagnoses based on what they consider the patient's most prominent symptom. It's a personal hierarchy developed through individual experience, the experiences of their teachers, and what they have read.

Sachdev focused on the intense intermittent pain that resolved quickly and left the patient completely normal in between. Her first thought was Crohn's disease, a form of inflammatory bowel disease in which the immune system wrongly attacks the gut. It usually presents in early adulthood. Celiac sprue was another possibility. Sprue, also known as gluten enteropathy, is an intolerance to gluten, a component of wheat. This patient did have a history of some unusual allergies, and sprue can cause transient episodes of stomach pain. To diagnose either of these diseases, Sachdev would need to pass a small camera through the patient's stomach and intestines and biopsy the characteristic findings. Finally, maybe there was still some scar tissue that was occasionally obstructing the digestive tract. She scheduled the patient for a barium test, which would allow her to watch the thick liquid pass through the small bowel. If there was any distortion because of scar tissue, it would show up here. The studies were completed over the next several weeks—and the results were normal.

At this point, the doctor told me, she knew she had to move off the beaten path. Whatever he had, it was going to be unusual. She thought about his strange allergies, the swelling of the hands and feet, which she had previously assumed were unrelated to his stomach problems. Could they be linked? Were all these symptoms caused by hereditary angioedema, a rare genetic disease that can cause allergy-like swelling? The swelling is usually found in the hands or feet but can also occur in the GI tract, which results in abdominal pain. Sachdev sent off blood to look for this unusual disorder.

It was three weeks before the results of the blood test came back. The patient did have hereditary angioedema. This genetic abnormality causes the immune system to be hyperresponsive, which results in localized swelling. The fluid that was picked up on the CT scans was a result of the same abnormality. While not well understood, the swelling attacks are thought to be triggered by injuries—even tiny ones—caused by anything from doing too many sit-ups to walking on hot sand to psychological stress. In this patient, it was possible that the swelling manifested itself primarily in the patient's gut because it was made vulnerable by his earlier appendicitis.

When she called the patient to tell him, he was silent for a moment. "Wait a second. If it's hereditary, how come my parents don't have this problem?" he asked. It was true—neither, it would turn out, had any evidence of this disease. The doctor explained that up to a quarter of all new cases of hereditary angioedema aren't inherited—they are new mutations. He was the first in his family to have the disease, but if he has children, he probably won't be the last. Each of his offspring would have a fifty-fifty chance of inheriting the disease from him.

Before returning to college, the patient was treated with anabolic steroids—the kind athletes are prohibited from using—which prevent this type of swelling, at least most of the time. I spoke with him later. He hadn't had an attack in more than a year. He said he feels great but worries about the future. "I'm not so sure about having kids. I don't think I'd want to pass this on. Not to anybody I loved."

Was It the Fish?

Dr. Kurtland Ma found the young man lying on the stretcher in the quiet of the predawn night. He was surprised by how healthy the patient looked: He had learned over the first year of his residency training that those who came to the Jacobi Medical Center emergency room in the Bronx at that hour were often the very sickest patients.

The thin chart reported that the patient came to the ER because he was having trouble walking. He had a headache; he felt weak and dizzy, and yet his vitals and initial blood work were completely normal. He was a puzzle, the senior resident told him as she handed Dr. Ma the chart. "I have no idea what's going on with this guy," she told Dr. Ma. "But he is probably going to need a head CT."

The patient was twenty-eight and said he was healthy until three days ago, when he and his girlfriend were in the Bahamas to celebrate his birthday. After a long day of swimming and snorkeling, they decided to try a restaurant they had heard good things about. They both ordered seafood—she had the red snapper, he the barracuda—and then went out dancing. Out on the dance floor the patient doubled over, caught off guard by an intense pain that knifed through his gut and took away his breath. He stumbled to the bathroom. The abdominal cramps and diarrhea came in waves. He kept thinking it would pass, but it didn't. Finally he decided to go back to the hotel.

As they walked through the streets crowded with other vacationers, his girlfriend teased him for letting a little bug ruin his birthday. But by then all he wanted was to lie down and go to sleep. Once in bed, sleep was elusive. His body ached with fever, and the cramping and diarrhea kept sending him to the bathroom. Finally he woke his girlfriend and told her he had to go to the hospital.

They were in the tiny emergency room in the Bahamian hospital when the vomiting started. Relentless heaves racked his body long after all he'd consumed had been eliminated. The rest of the night was a blur of tests and treatments punctuated by slowly diminishing waves of pain and nausea. The Bahamian doctors returned to the patient's bedside frequently. He was feverish and even the lightest pressure on his abdomen was excruciating. Was this appendicitis? Hepatitis? Or just a bad case of food poisoning? The CT scan showed a normal appendix. The blood tests showed no signs of hepatitis or any other infection. The antinausea medication stopped the vomiting and slowed the diarrhea.

This was probably food poisoning, a doctor told the exhausted patient. Most food poisoning is caused by ingested bacteria—*E. coli*, salmonella, or *Staphylococcus aureus*. Seafood-related food poisoning is often linked to a less well-known bug—*Vibrio parahaemolyticus*, although that bacteria is usually killed by cooking. Had they eaten sushi? No, their food was well cooked, the girlfriend assured him. The doctor shrugged. Generally it doesn't make sense to try to identify the bug, because the treatment tends to be the same no matter what you've got, he told the couple. The most important thing was to avoid becoming dehydrated, and this patient was getting plenty of fluids.

By morning, the patient felt a little better. He was given a prescription for an antibiotic and something for the nausea and sent back to recover in his hotel. He slept for the next two days. Finally the patient felt well enough to venture out. As he dressed he noticed that his hands seemed clumsy. And his feet felt as if they were asleep, as if he were walking on a shifting carpet of tiny nails.

He wasn't sure he could eat. His girlfriend bought him a smoothie from a juice stand. The fruit drink smelled delicious, and his stomach rumbled eagerly. He took a sip and immediately spat it out. The icy cold drink felt as if it had come straight off the stove—as if it were boiling hot rather than freezing cold. He took another sip. His mouth burned. It felt too hot to swallow. At this point, the patient decided he'd had enough, and the couple soon flew back to New York. The young man dropped off his girlfriend at home and continued on to the Jacobi emergency room.

Dr. Ma was taking notes as the patient told his story, but when he mentioned this strange reversal of hot and cold, the doctor gasped. "I know what this is!" he shouted, interrupting the patient's story. "I know what this is!" And with that he ran down the crowded hallway to where the attending physician and senior resident were sitting. "He doesn't need a head CT! He has ciguatera poisoning."

Ciguatera poisoning comes from eating fish that has

been contaminated with a toxin produced by an organism that grows on reef algae in some infested tropical waters. Because the toxin is stored in fat, its concentration increases as it moves up the food chain from the little fish who eat the tainted algae to the larger, predatory fish, like shark, snapper, grouper, and barracuda, and from there to the human consumer. Unlike most other causes of food poisoning, this toxin is colorless and odorless and isn't destroyed by cooking.

The illness was first described in 1774 by a surgeon's mate on the crew of Captain Cook's South Pacific exploration aboard the HMS *Resolution*. The crewman, John Anderson, documented the symptoms described by several shipmates who had eaten a large fish caught in the tropical waters. There was "a flushing heat and violent pains in the face and head, with a giddiness and increase in weakness; also a pain, or as they expressed it, a burning heat in the mouth and throat." Many since then have described the rapid onset of nausea, vomiting, and diarrhea—similar to other types of food poisoning—but followed by the kind of strange neurological symptoms this patient had. Alterations in sensation—like the numbness, tingling, and bizarre hot-cold reversal—are most common and most characteristic. The toxin can sometimes affect the heart—causing it to beat too slowly or irregularly. It is rarely fatal, but there is no effective treatment, and the symptoms can persist for weeks, sometimes months, occasionally years.

It was a great diagnosis, the senior doctor told Dr. Ma. But how did he know? It was easy, Dr. Ma told his doctor-teachers. He took care of a family with ciguatera poisoning several months earlier. A whole family had eaten barracuda for their Christmas dinner. They came to the hospital a few hours later, after the nausea, vomiting, and diarrhea had given way to these strange neurological problems. He'd never forget them.

Dr. Ma went back to the patient's room. He apologized for his unexpected exit and began to explain the illness and where it had come from. Although the pathology of this poison is still not well understood, current thinking is that the toxin damages the protective sheath covering the nerves, causing the sheath to swell and compress the delicate tissue it's supposed to protect.

"Even before they told me, I knew it had to be the fish—the barracuda," the patient told me sadly. Barracuda has recently been identified as a common source of the toxin, and the CDC now advises against eating the fish, especially when it's been caught in the Caribbean. These days ciguatera is not just a tropical threat. In the United States, it has become one of the most common fish-related illnesses as the waters off the coasts of Florida, Texas, South Carolina, and most recently North Carolina become warm enough to host these once tropical organisms.

More than six months after the patient returned from the Caribbean, he still had not fully recovered. He could eat again—after losing twenty pounds in the first few weeks of the illness. He still had occasional numbness and weakness. The patient sighed: "And it was such good fish, too. I ate a lot of it."

A Bad Stomach Gets Worse

The stretcher carrying the nineteen-year-old woman was hurried into the emergency department of Banner–University Medical Center in Tucson, Arizona. Her breath moved in and out erratically. Her jaw was clenched shut. She held her arms stiffly at her side.

The night before, the girl had called her mother and told her she was vomiting and had a terrible pain in her stomach. Her mother, who lived in Texas, asked if she needed to go to the emergency room. The girl thought not, so the mother offered to make a doctor's appointment for her the next day. But when her mother called her in the morning, to let her know the time of the appointment, she got no answer. Worried, she called her older daughter, also a student at the University of Arizona, in Tucson, and asked her to check on the younger. The sister found her in her dorm bathroom, unconscious and covered in vomit. The water was still running in the sink, and an electric toothbrush buzzed nearby, suggesting a sudden, unexpected collapse.

The young woman had no medical problems, her sister told the emergency team. But she had gone to the ER three months earlier—taken by friends who noticed that she seemed confused. At that visit, she had an MRI of her brain and some blood tests. The tests were unremarkable except for a minor abnormality in the test of her liver function. (She saw a doctor for that a few weeks later who didn't pursue it when her levels normalized.) She got better with time and some IV fluids and was sent home later that night. Though testing showed no alcohol or drugs in her system, some of the doctors wondered if she'd taken something—maybe synthetic marijuana or LSD—that wouldn't show up on their tests.

In the ER, the comatose young woman reacted to pain, but nothing else. She remained rigid, from her jaw to her feet, which suggested a terrible brain injury. But a CT scan of her brain looked normal. The blood tests were likewise unremarkable, save for the same minor abnormality in her liver-function test. Her heart was racing, and her breathing was ragged and irregular enough that the ER doctors put her on a ventilator, a machine that breathed for her, to make sure she got enough oxygen. She was sent to the intensive care unit and given broad-spectrum antibiotics and an antiviral, in case whatever she had was caused by an infection.

These are perhaps the toughest cases that doctors face: A patient comes in critically ill—dying, really—with few clues as to what is going on. The ICU doctors examined the young woman as soon as she arrived but had no better understanding of what was causing her coma. They asked for help from neurology, cardiology, and infectious diseases. Those doctors were baffled as well. An MRI revealed that the girl's brain had swollen and was now pushing up against the unyielding limits of the skull. If the cause of the swelling was not found and reversed, she would certainly die.

An EEG (electroencephalogram) suggested that the patient's rigid posture might be a result of an ongoing seizure. The patient was started on antiseizure medicines. Although the seizures stopped, the patient didn't wake up. Indeed, it was clear that she was getting worse. Her eyes, initially reactive to light, became fixed and dilated. And when icy water was squirted into her ear—a stimulation that would normally cause a strong involuntary reaction—nothing happened, indicating that her brain was not working, even at the most basic level.

The team stopped all potentially sedating medications to see if they were contributing to the problem. There was no improvement. And so, after a thorough work-up and thoughtful care, eight days after this young woman was brought to the hospital, she was taken off the breathing machine. Without its assistance, she breathed no more, and the team pronounced her brain-dead.

Why did this happen? The family asked over and over, but the doctors who had cared for the young woman couldn't tell them. Her brain had swollen, and that swelling killed her, but beyond these basic facts they simply didn't have an explanation.

The family, though grieving, decided to donate the young woman's organs. Her heart, liver, and kidneys found grateful recipients. Then the family took their child's ashes home and made plans to scatter them in a beloved place.

A few weeks after the young woman's death, her father got a call from the Donor Network of Arizona. The recipient of his daughter's liver had died, too, just days after the transplant surgery. Shocked, the transplant team immediately went to work to find out why. The answer was a surprise. Their daughter, the donor of the liver, had been born with a rare genetic defect. She was missing the genes that made a chemical called ornithine transcarbamylase (OTC)—a vitally important enzyme whose job is to help the liver break down proteins. Proteins are the building blocks of muscles, and something as simple as eating meat, or stressing the body through, say, fasting or surgery, can release additional proteins into the system. When something goes wrong with the process that breaks down protein, toxic levels of ammonia can build up, which in turn attacks the nervous system and the brain.

Hyperammonemia, the medical word for high levels of ammonia, is usually caused by a liver so damaged by alcohol or disease that it can't perform even its most basic jobs—including the breakdown of proteins. Doctors routinely test ammonia levels in patients who have liver failure, often from cirrhosis or hepatitis. But patients with OTC deficiency usually don't have a bad liver; theirs simply fails to do this one job. And because the organ is working fine in all other ways, doctors don't suspect hyperammonemia as a cause of coma or brain swelling. If this young woman's ammonia level had been checked, it would have been high—maybe ten times the normal level.

OTC deficiency is not the only unusual cause of elevated ammonia. There are other diseases—some inherited, like OTC deficiency; others acquired, like herpes and certain cancers—that can do it as well. Medications, including iron and antiseizure drugs, can also cause elevated ammonia levels. A simple blood test, performed when no other cause of coma is found, may save a life. And in the case of this young woman, possibly two.

The gene for OTC is on the X chromosome, and so boys, who have only one X chromosome, are more likely to have symptoms if they have this genetic anomaly. Because girls have two X chromosomes, if one is defective, the other can often make up for it. But for reasons that are not well understood, in the right setting, maybe after a protein-rich meal or with significant stress, like an illness, the condition can manifest in girls, too, prompting their ammonia levels to spike upward. That's probably what happened to this patient when she got sick and confused three months earlier. And that is what killed her and her liver's recipient.

As the parents read up on OTC, they recognized several symptoms in their daughter. Like many with OTC deficiency, she had always had a bad stomach—feeling nauseated or vomiting for reasons they never understood. And she rarely ate meat—she'd never liked it.

The rest of the family was tested for OTC deficiency, and the girl's father found that he, too, had the defective gene. This discovery explained so much to him—why, for example, there were times when he literally felt too tired to move or even speak. That lethargy, he now believes, is a response to a high ammonia level. These days he takes a supplement that provides the chemicals his enzyme can't produce. And he avoids the foods that can raise his ammonia.

All this information came too late to help their daughter. But perhaps others will find it useful. That, at least, is her parents' prayer.

A Hockey Stick to the Gut

The doctor found his twenty-year-old son in the bathroom sprawled over the toilet. "Not again?" he asked gently. The young man nodded, tears bright in his eyes, as he rose slowly to his feet. He pressed his hand deeply into his own abdomen, as if holding something in place. "It's getting worse."

The father was overwhelmed by a sense of helplessness. "Get dressed," he told his son suddenly. If they rushed to the hospital, maybe they would be lucky enough to catch whatever was causing this pain on an X-ray. The young man had already been imaged a half dozen times, but never during an attack. But a short time later, as they walked down a quiet hospital hallway, he turned to his father. "I'm sorry, Dad," he said. "The pain is gone." As it had so often in the past, the attack ended the way it started—suddenly. The X-ray was normal.

The young man's father, a gastroenterologist, had been

trying to figure out the cause of these terrible episodes for months. He was tormented by the possibility that he was missing something. It was, he thought, time to send his son to another doctor, and so he called an old friend and internist, Andrew Israel.

Israel was shocked by how much weight the young man had lost since he'd last seen him. As he hugged him he could feel the bony knobs of his spine beneath his thin shirt. The young man began to describe the strange pain that had come from nowhere and dominated his life for the past three months. It was a tearing, burning pain, always in the upper-left side of his abdomen. And it would come on suddenly—often just after he ate. These excruciating attacks would last a few hours, then, just as unexpectedly, disappear, as if nothing had happened. Recently the attacks were coming more frequently and lasting longer, and were often accompanied by nausea and vomiting. He tried not to eat, since that seemed to be one trigger, but even that wasn't working anymore.

The young man paused, then added that just a few days before his first episode of pain, he had gotten injured playing hockey—a violent collision on the ice where his stick was jammed hard into his rib cage. The patient said he felt that the two were related even though the stick hit him on the right side and this strange, intermittent pain was on the left. He had no other pertinent medical history. He took no medicines, didn't smoke, drank occasionally, was very physically active. He had already been to several specialists and had lots of tests: blood work, a CT scan, several MRIs. Nothing panned out.

On exam, the patient's abdomen was soft, with normal sounds. Israel could feel no masses, but the pulsation of the aorta was surprisingly prominent. Was it pronounced because the boy was so thin? Or was there something wrong with the muscular pipe that carried the blood from the heart to the rest of the body? Could this be Marfan syndrome, a congenital abnormality of the connective tissues of the body? The boy had the body type of those who carry this genetic mutation—tall and thin with long arms and legs. In this disease, the defective tissues can't hold up against the onslaught of pressure in the aorta, and patients frequently develop areas of weakness that are prone to rupture. It seemed very unlikely, but it would be tragic to miss.

What else then? These words are the most essential component of a difficult diagnosis. What else could this be? The pain started after the hockey accident—but Israel could see no way to link the injury on the right with this recurring pain on the left. Obstruction in the small intestine could cause pain and vomiting after eating. But why would he have an obstruction? In adults, an obstruction is usually the result of the scar tissue that forms after an operation, which prevents the small intestine from moving normally, but this young man had had no operations. Kidney stones could cause this kind of intense, intermittent pain. But that should have been picked up on the CT scan he already had.

Israel told the young man that he wasn't sure what was causing his pain. He suggested a second CT scan of his abdomen. He'd need to look over the other tests before he could figure out what else might be needed. It pained the internist to see the hope fade away from the boy's face as yet another doctor disappointed him.

The next day, Israel pored over the results from the various studies. The CT scan was normal; MRIs—normal. A barium study of the stomach and intestines was also normal. The blood work showed no evidence of any infection or inflammatory disease. If a second CT scan didn't show anything, he wasn't sure what else there was to do.

The CT scan was scheduled to be done a few days later. As the patient drove to the test, he felt the start of the now-familiar tearing and burning and nausea. But this time, the dread of the hours of pain was matched by a cautious optimism. Maybe, finally, they would see what was causing these attacks if they scanned him while he was still in pain. The radiologist watched the images. The aorta was fine. But what was that? A portion of the small intestine looked bulkier than the rest. The walls were thicker than normal, the inner lumen almost completely obliterated. It was an intussusception—an unusual condition, in which one part of the small intestine is telescoped into an adjacent part. This could be a surgical emergency. The abnormal folding of the intestinal tube can obstruct the flow of blood. When that occurs, the bowels can die. In previous attacks, when the intestine folded, it must have unfolded on its own, relieving the pain and restoring the blood flow. But with each attack, the delicate tissues of the small bowel were repeatedly injured, becoming swollen and inflamed, making it more likely to telescope again and more difficult to unfold.

In adults, the most worrisome cause of intussusception is cancer. The presence of a mass inhibits the normal movement of the bowel, making this collapse more likely. They wouldn't know what exactly caused this intussusception until they operated.

In the OR, when the surgeon removed the damaged segment of small bowel, the cause of the problem became clear: The patient had what is known as a Meckel's diverticulum. This is a common congenital abnormality—a remnant of a duct that normally provides nutrition to the embryo. In utero, it links what will become the umbilical cord to a primitive gastrointestinal tract and usually disappears before the fetus enters the second trimester. In the patient, this pinkie-size flap of tissue never went away, and it somehow interfered with the normal movement of the bowel, leading to the telescoping of the small bowel into itself.

The usual teaching about Meckel's diverticulum is the rule of twos: it occurs in 2 percent of the population, but only 2 percent of those with a Meckel's will ever have any complications; the vast majority of those who do will be under the age of two. No one knows why this common abnormality, usually so well tolerated, can suddenly begin to cause problems. In this patient, perhaps the hockey collision injured the small intestine—perhaps the Meckel's tissue itself—in a way that set these attacks off.

In a difficult case, close observation, careful logic, and stringent deduction will often lead to a proper diagnosis. But there are times when the doctor must rely on the progression of the disease to provide an opportunity to unravel the mystery. Patients with intussusception caused by a Meckel's diverticulum can suffer for months, even years, before a diagnosis is made. The disease is too unusual to reason out and too transient to count on catching until the damage to the gut is advanced and the condition becomes life-threatening, as it was in this young man. I've often heard people say that they'd rather be lucky than right. In a rare disease like this, you really need to be both.

Age of Embarrassment

A wave of nausea swept over the elderly woman as she climbed the stairs of her daughter's home. She hadn't felt well since she left her own tidy studio apartment in Florida the day before. She lowered herself onto the stairs with a quiet moan; her face was pale, her coral pink lipstick faded. "I'm so sorry, darling," she said as her daughter ran into the hallway, "I threw up on your stairs." Embarrassed, she admitted that she hadn't been able to keep anything down since the morning before. She'd thought about not visiting, but at ninety-three, how many more Christmases would she be able to spend with her daughter and grandchildren? "I kept thinking it was going to go away, but it just hasn't," she explained.

"Mother, really!" the younger woman chided gently as she helped her into her bed and briskly checked her blood pressure. It was high: 200 over 80 (normal is less than 120 over 80). She immediately called her mother's doctor in Florida, who told her to take her mother to the emergency room.

That evening, Dr. Ben Musher listened as Radhika Varada, a third-year resident, briefed him on their patient: a ninety-three-year-old with a history of high blood pressure, emphysema, and kidney cancer who had come to the ER after two days of nausea, vomiting, and lethargy. Varada reviewed the data collected by the ER physicians, then brought Musher to the patient. He noted that she looked much younger than her considerable years. He briefly recounted what he knew of her story. She had nothing to add.

The two doctors examined the patient together—Varada had already done so when the patient was admitted. Her blood pressure was still high—though not as high as it had been at home. Her abdomen was soft and a little sore from the effort of vomiting, but otherwise unremarkable. A CT of her abdomen hadn't shown anything except that she was missing her left kidney, which had been removed when her cancer was diagnosed four years before. A chest X-ray was also normal. The most interesting finding came from some routine blood work: the patient's level of sodium—an essential component of blood chemistry—was dangerously low.

The doctors in the emergency room had attributed the patient's low sodium, or hyponatremia, to her days of vomiting and subsequent dehydration. They had started her on a slow infusion of intravenous fluids. Certainly dehydration is one of the most common causes of hyponatremia—especially in a patient like this who has been vomiting or had diarrhea. A reasonable hypothesis, but Musher didn't think it was right. The physical exam didn't support a diagnosis of dehydration: Her blood pressure was high; her heart rate was normal—it's normally rapid with dehydration. And her urine was dilute—it should be highly concentrated if she had too little water in her system. Given this evidence, Musher thought the vomiting was caused by the low sodium rather than the other way around. But then what caused the low sodium?

Musher focused on the most likely possibilities for the elderly woman. First and foremost: medications. Many common drugs can cause low sodium, and this patient was on lots of medications. The two doctors carefully went over the neatly handwritten list of medicines the patient carried with her, but none were known to cause low sodium. Some unusual diseases of the hormone systems can cause hyponatremia: Addison's disease—which occurs when the adrenal glands stop producing their hormones—can cause the body to lose sodium. Too little thyroid hormone can decrease sodium levels as well. A simple blood test would show if either of these was the cause. But what worried Musher most was the possibility of cancer. The patient's extensive smoking history put her at risk for lung cancer, which can cause hyponatremia. The cancerous cells produce a hormone that mimics one made by the body to regulate water. Too much of this hormone (called vasopressin) will cause the kidneys to hold on to water. They'd need more studies to make a diagnosis. In the meantime, they stopped the fluids and instructed the patient to limit her water intake in order to give her remaining kidney a chance to restore the proper balance of salt and water.

On rounds early the next morning, the two doctors stopped to see their nonagenarian. She felt much better, she said. And she looked it: her white hair was combed to reveal a stylish cut, and her lipstick was newly applied. Her sodium was improving, though it was still far from normal. The other blood tests—to check the thyroid and adrenal glands—were normal. As he considered what to do next, the patient's daughter approached him. Her mother looked much better, said she felt much better—was she better? It was Christmas Eve. Could she possibly go home to spend the holiday with her family?

Musher hesitated. At this point, he felt that the most likely cause of her hyponatremia was cancer. She needed the work-up. On the other hand, it was Christmas—if she stayed, would anything really get done?

After instructing the daughter about the fluid restriction, Musher allowed the patient to go home. "If it were lung cancer, there was a good chance this really would be her last Christmas with her family," he explained to me. He encouraged the patient to follow up with her doctors when she returned to Florida; they would need to determine the cause of her symptoms.

Four days later, mother and daughter were back in the emergency room. Christmas had been good, but now the patient was feeling lousy again. Her sodium was better but still very low. Dr. Varada greeted them with a smile. "Let's see if we can figure it out this time." She pored over the data collected by the emergency room. Then she turned her attention once again to the list of medicines. "Are there any other medicines, or herbs or over-the-counter drugs—anything at all that you're taking that is not on this list?" she asked once more. The elderly woman thought for a moment. "Well, I don't know if the new medicine is on that list," she offered tentatively. She didn't know the name of the drug, but her urologist had given it to her so she wouldn't have to get up to go to the bathroom three, four, five times a night. She had tried the pill, hadn't liked the way it made her feel, and stopped taking it. But she decided to try it again on this trip; it would just be simpler, she explained, if she didn't have to get up again and again in her daughter's home.

The patient looked away. Her daughter hadn't known about this problem, but she wasn't really surprised her mother hadn't told her. Her mother always worked so hard to hide as many of the effects of aging as she could—even from her own daughter. She immediately called her husband at home. A comparison between the medicines in her mother's travel bag and those on her list turned up the culprit. The label on the bottle said DDAVP. It was the synthetic form of vasopressin—the hormone that causes the body to hold on to water. In dealing with this very capable elderly patient, the doctors had forgotten a basic principle in geriatrics—when it comes to medicines, trust but verify.

I spoke with the patient later about her Christmas medical emergency. She could barely remember it. "The whole thing seems like a dream," she told me, a reflection, most likely, of the significant effect the electrolyte imbalance had on her brain. And it took her weeks to feel normal again. She was annoyed with herself for not having put the drug on her list, but equally annoyed with the doctor who prescribed it without telling her about this very common side effect. But then she just shrugged it off. "Oh, you know, it's just getting old," she told me. "I didn't know it was going to be like this. You feel like you're the same person—but you're not. And when you forget that, nothing goes right."

Hurts So Bad

The scream tore through the dark apartment. "Mommy, Mommy, Mommy!" The woman leaped out of bed and hurried down the hall to her daughter's room. She moved quickly and quietly to the nine-year-old girl lying still under the flowery quilt.

"It hurts, Mommy. It hurts so bad."

"I know, sweetie. I know." The mother knelt carefully next to the bed, making sure she didn't jostle the mattress. Experience had taught her that any movement made a terrible pain even worse. As she stroked her daughter's mass of curls, she could feel the warmth of her skin and knew she had a fever. In a few minutes, she would get the child some Tylenol and a hot towel to hold against her rioting stomach. She wasn't sure if these rituals helped, but they gave her something to do to combat the helpless despair of watching her baby suffer this way over and over again.

Her daughter could always tell when an attack was brewing. After supper that evening, she had sought out her mother. "It's coming," she whispered as she wedged herself into her mother's lap, her face pale, her lips almost white. The mother put her daughter to bed, and she and her husband waited. Maybe this time it would be different. But it wasn't.

These mysterious episodes of fever and stomach pain had started two years earlier. It was their older son's ninth birthday party, and their daughter had just started first grade. That first attack the mother attributed to excitement or anxiety or maybe something she had picked up at school. Her daughter had been well enough to eat some birthday cake, and she thought nothing of it until the same thing happened three weeks later. For the next year, every three or four weeks her little girl would get a fever and these strange and terrible stomachaches. She wouldn't eat. She would barely drink.

The pain was usually on the right side. The child variously described it as sharp or crampy or achy. Sometimes she felt nauseated. She would lie curled on her side, her breath the only visible movement. After an hour, sometimes two or three, the pain would seem to ease up, and she'd slide into sleep. In a day or two, she would be back to normal—until the next attack.

Her pediatrician was baffled and sent the child to a gastroenterologist. When he wasn't able to figure it out, her mother found Dr. Joseph Levy, the head of pediatric gastroenterology at New York University School of Medicine. He looked into her stomach and intestines with an endoscope in search of the cause of the painful episodes. Was it celiac disease? Ulcers? Crohn's? The studies said no.

The only abnormal test result was an elevated sedimentation rate. This test measures inflammation based on how quickly red blood cells sink to the bottom of a test tube. The test suggested inflammation but couldn't reveal where or why. Levy tested the patient for lupus, the most common chronic inflammatory disease among girls this age. The test was normal. Levy called the girl his little mystery and continued to search for an answer.

Her mother conducted her own investigation. She trolled the Internet and described her daughter's symptoms to every doctor she knew. At one point, she outlined the symptoms to her own internist, and that finally provided some hope. After hearing about the regular bouts of fever and pain, he said right away, "Sounds like familial Mediterranean fever." He had never seen the disease but recognized the pattern from medical school. The mother quickly called Levy.

The specialist knew the disease well, he told her. It's a genetic disorder usually seen in ethnic groups from the Mediterranean region. He had trained in Israel, where the disease was common, and her daughter's symptoms didn't fit the disease he knew. These children, he told her, have high fevers and an abdomen that's rigid with pain. The disease is often mistaken for appendicitis, and many children end up in an operating room before their doctors figure out the diagnosis. He had examined her during an episode, and while he wasn't sure what she had, he was pretty sure it wasn't FMF. The mother could feel tears prickling her eyes. There had to be an answer somewhere.

That winter, the patient continued to have attacks of abdominal pain every few weeks. Then she developed pain in her right ankle. A second test for lupus was now abnormal. Based on this, Levy referred the girl to yet one more doctor: this time, a pediatric rheumatologist, Dr. Lisa Imundo.

In her office, parents and daughter retold their story once more as Imundo took notes. She asked about any other joint pains. Sure, there had been other aches and pains—mostly in the knees—but for the past few weeks the pain had moved to the ankle. The mother hadn't mentioned it to Dr. Levy because her daughter played sports, and she had assumed the pains were due to little injuries. Any tick exposures? Imundo continued. Yes, they had a house in an area known to have deer ticks.

On exam, Imundo noted that the patient was a little overweight and more than a little anxious. Her abdomen was soft, her bowel sounds normal and she had no stomach tenderness. Her ankle was sore and had a limited range of motion but wasn't red or swollen. Finally, Imundo laid out her plan. Since the second test suggested she might have lupus, she would send off blood to see if there was any other evidence that she had this puzzling autoimmune disease. It wasn't a classic presentation, but the symptoms of lupus were protean. Lyme disease was also possible, though less likely. The stomach pains weren't typical, but the wandering joint pain was.

"What about familial Mediterranean fever?" the mother asked about her internist's suggestion. It's true that the disease was characterized by abdominal pain, Imundo reasoned. And patients with this disease were completely normal between episodes of fever and pain. But the abnormal blood work, taken when this child was healthy, would be unusual. However, a test had recently been developed for this unusual genetic disorder, she told them, and they could check.

A week after the blood was drawn, Levy called the mother. "I was wrong," he told her straightaway. "Your daughter has familial Mediterranean fever." She and her husband each must have one copy of the FMF mutation, he explained. Carriers, with only one copy, have no symptoms but can pass it on to their children. With two copies, the body makes an abnormal version of a protein called pyrin, which is essential to modulating the immune system. Because of this malformed protein, an army of white cells that would normally protect the body somehow overreact, resulting in inflammation, pain, and fever. The abdomen and joints are the most common sites of these attacks, but the lungs and heart can also be affected. Colchicine, a drug that inhibits some forms of inflammation, can prevent most attacks.

They started their daughter on colchicine the next day, and the attacks stopped as abruptly as they had started. The nightmare was finally over. As long as her daughter takes her medicine, the mother told me, the fever and pain are kept at bay.

I asked Levy why he had been so certain that the child didn't have FMF. Until recently, he explained, FMF had been a clinical diagnosis—one made based on the patient's symptoms and the physical exam. The identification of the gene and the subsequent development of the test—which had only just become available at the time of this patient's diagnosis—changed doctors' understanding of the disease, Levy told me. In medicine, we can only really know a disease once we have a test that can reliably identify it. "What we learned from the test was that there was a whole spectrum of disease," Levy said. "Before, we were only able to pick up what we now know were the most extreme forms of the disease—the tip of the iceberg. Now we can find all the rest."

Knifed

"I'm dying," the fifty-seven-year-old woman whispered. Though they sat right next to her bed, her two sisters had to strain to hear the words. "I can feel the life just slipping away from me." They were going to take her to the hospital, the oldest sister told her. But not to the same hospital she'd been to so many times already.

The patient hadn't been healthy for years. Her rheumatoid arthritis was bad enough to force her retirement from nursing a decade ago, and nothing she took seemed to help. But the joint pain was at least endurable. For the past couple of years, though, she had a new pain—one much harder to bear. Every time she ate, it was like being stabbed with a knife in her belly. She didn't have health insurance, so she'd tried to figure it out for herself.

The pain came on whenever she ate, but it seemed worse when she ate bread or pasta. A diagnosis of celiac disease totally fit: In this disorder, gluten, a protein found in grains like wheat and rye, triggers the body's immune system to attack the absorptive lining of the gastrointestinal tract. Consuming gluten-containing foods destroys the gut's ability to absorb nutrients, causing pain, diarrhea, and malnutrition. And yet, when she tried to avoid gluten—not easy, because it seemed to be in just about everything—the pain still didn't go away.

Within a few months, her stomach hurt all the time. The sharp pain still came after she ate, but in between there was this dull ache, as if she were recovering from a punch in the gut. And no matter how careful she was with her diet, she suffered diarrhea a dozen times a day, every day. She lost more than fifty pounds that first year.

After months of this, she was so weak she could barely walk. She had to go to the hospital, insured or not. In the emergency room, the cause of her weakness was quickly identified: She was severely anemic, and an essential electrolyte, potassium, was dangerously low. Potassium is necessary for muscle cells to do their work.

Once she was given the needed blood and potassium, the doctors in the ER set out to determine why she had these deficiencies in the first place. The retired nurse told them her celiac theory—which, they thought, made sense.

Untreated celiac can cause profound nutritional deficiencies. A gastrointestinal doctor scoped her stomach and duodenum. The villi, the tiny fingerlike tentacles that normally cover the inner surface of the intestines and absorb most of the nutrients, had been flattened. The most common cause of this kind of intestinal destruction is celiac disease.

Her doctor instructed her to avoid all gluten. The patient told him how frustrating that was to hear, given how hard she had been working to avoid it. Yes, the gastroenterologist acknowledged, it's difficult to eat this way. But it was the only way to prevent this kind of devastating damage.

And yet, no matter how she shaped her diet, the symptoms persisted. The doctors accused her of cheating on her diet. "Noncompliant" was the word they used. She couldn't imagine where the gluten could be coming from. She ate only meat and vegetables. Yet the pain and the diarrhea continued. Her weight dropped to seventy-five pounds. She hardly recognized herself when she struggled past the bathroom mirror on her way to the toilet.

At her last trip to the hospital, the doctors told her she'd had a heart attack and possibly a stroke. There was nothing they could do for her if she didn't work harder on her diet. So she went home. If she was going to die, she didn't want to do it at that hospital. She lay on the couch, too weak to do anything for herself. Her son and her sisters took turns cleaning her, dressing her, and feeding her. She had to be lifted onto the bedside commode. She ate only mashed potatoes. She was dying.

It was early morning when her sisters convinced her to let them bring her to the University of South Alabama Medical Center in Mobile, just an hour away. Dr. Heather Fishel, a resident in her second year of training, introduced herself while the patient was still in the ER. The woman in the bed looked much older than fifty-seven. The skin on her face was so thin and pale you could almost see the bones beneath.

The ER doctor had already determined that, once again, the patient had severe nutritional deficiencies, so even before Fishel arrived, she was getting intravenous fluids and potassium. As Fishel tried to get more of her story, the patient became impatient: "I've already told this to the ER doctor. Don't y'all talk to each other?" All the patient would tell her was that she had celiac disease and that it was killing her even though she was on a gluten-free diet.

Fishel sent off the patient's blood to look for the antibodies that prove the autoimmune reaction of celiac. The results confirmed what she suspected: no antibodies, which meant no celiac. The patient had a second biopsy of her stomach and duodenum. Like the first, it showed destruction of the villi. But knowing now that she didn't have celiac, they had to look for a different cause of that devastation.

Dr. Leonel Maldonado was the pathology resident on duty when the patient's biopsy came to the lab. He, too, noted the flattened villi so suggestive of celiac. But he noticed something else as well. In the layer just below the surface, he could see cells that didn't belong there. These were macrophages, the police vans of the immune system—whose job it is to capture invading bacteria, cart them off, and destroy them. These white blood cells were bulging with something that they hadn't been able to eliminate. Was it tuberculosis? Detritus from an unseen cancer? Or was it some kind of bacteria? And why weren't these captives being destroyed? Maldonado worked to unravel the mystery, one test at a time.

After two weeks, the patient was well enough to go home, but she still had no diagnosis. And she was starving. She discovered that if she didn't eat, she didn't have the diarrhea. So she refused to eat.

But just two days after her discharge, Maldonado had the answer. The stuff inside the macrophages was a strange bacterium: She had something called Whipple's disease. This disorder was first described by George Whipple in 1907. He was caring for a fellow physician who had terrible weight loss, diarrhea, and arthritis. When the man died, Whipple noted at autopsy the foamy macrophages, which, as with this patient, were filled with a bacterium that was later named *Tropheryma whipplei* (from the Greek *trophe*, meaning "nourishment," and *eryma*, or "barrier," a reference to the nutrient malabsorption that is characteristic of the disease).

This bacterium lives in the soil just about everywhere yet rarely causes disease. Up to 70 percent of the healthy population has antibodies to it, suggesting that infection is usually successfully fought off by most. Those who develop Whipple's are thought to have some immune-system defect that the organism can take advantage of. When captured by macrophages, the bug somehow disarms the cells' built-in mechanism for destroying their prey.

The treatment for this infection is a year of antibiotics. As soon as the team heard the result, they called the patient's house. She wasn't there, her son told them—she was at the hospital, seeing a cardiologist for follow-up on her heart issues. They found her, admitted her, and started treatment. After the first couple of doses of antibiotics, her appetite came back. Within days she was able to eat without getting sick.

The road back was long and difficult. She was so debilitated that it took her more than a year to get to what she considers a normal life. But finally she was able to give back the borrowed wheelchair and put away the walker. She can walk without a cane, but she still can't go far. She is amazed by how close she came to dying and wonders if it would have taken so long if she hadn't offered her own diagnosis.

They tell me it's a rare disease, she said when I last called her. But is it really rare, she wondered aloud, or is it just rarely looked for?

It's a good question.

Suddenly Sick, Again

"Mommy, I'm afraid. Tell me what to do." The child's mother looked up at her eight-year-old daughter. "It's going to be okay," she said. "Just go get some help."

The woman watched as her daughter left the public bathroom, where she now lay. She and her daughter had come to this store to pick up some new towels. But once inside the mother began to feel hot and dizzy. Her heart fluttered in her chest, and she felt as if she was going to be sick. She grabbed her daughter's hand and hurried to the bathroom. Once there she suddenly felt as if she was going to pass out and laid down on the bathroom floor. That's when she sent her daughter to get help.

Finally, a store clerk came into the bathroom holding the little girl's hand. The last thing the woman remembered was the look of horror on the clerk's face as she saw the middle-aged woman lying on the floor in a pool of bloody stool.

When the EMTs arrived at the store, the woman was

unconscious. Her heart was racing, and her blood pressure was terrifyingly low. She was rushed to the emergency department of Yale New Haven Hospital.

But by the time she arrived there, her blood pressure had come up and heart rate gone down, and she was no longer bleeding from her rectum. A physical exam uncovered nothing unusual, and all of the testing she had was normal, with one important exception: her blood seemed to have lost its ability to clot. If that problem persisted, she would be in danger of bleeding to death after even the smallest cut or abrasion.

The patient told the ER doctors that her only medical problem was anxiety that caused occasional panic attacks, and she had recently started taking an antidepressant for that. She didn't smoke, rarely drank, worked in an office, and was married with two children. She had been healthy her whole life until almost two years before, when the exact same thing happened to her: One day, out of nowhere, she had sudden, bloody diarrhea, her blood pressure dropped, and she lost consciousness. Then, when she got to the hospital, doctors found that her blood would not clot.

Dr. Susanne Lagarde, a gastroenterologist, was asked by the medical team to see the patient to help figure out why she had bled from her gut. Lagarde introduced herself and then quickly reviewed the events leading up to the incident in the store. But she also wanted to know the details of the last time this happened to the patient. Did the doctors ever figure out why her blood didn't clot? No, the patient said, they couldn't figure it out in the emergency room, and the following week when she saw a hematologist—a specialist in disorders of the blood—her blood was completely normal.

Lagarde recommended a colonoscopy, a procedure that

uses a small camera to look at the tissue of the large intestine, in order to determine why the patient bled. The most common cause of bloody diarrhea is inflammation of the delicate tissue of the large intestine. This can be caused by an infection or diseases like ulcerative colitis or Crohn's, autoimmune disorders in which the white blood cells that are supposed to protect the body from invading pathogens mistakenly attack completely normal cells.

But when Lagarde looked through her scope she saw none of that. The delicate lining of the colon was damaged in several places, but it looked as if that had occurred when the gut cells were injured by the lack of oxygen-carrying blood caused by the same low blood pressure that made the patient lose consciousness. So this wasn't a problem of the gastrointestinal tract. The inability to clot turned a tiny trickle of blood from the injured tissue into a torrent. So what caused that combination of hypotension and difficulty in clotting? Certain severe infections can cause both. But there was nothing to suggest that she had an infection. There is a medication—heparin—that can cause a brief period of anticoagulation. Heparin is an intravenous drug that is used to treat patients who develop harmful clots. Intentional misuse of the drug seemed unlikely, and Lagarde couldn't imagine any kind of accidental exposure. But one thing seemed clear: this patient needed a diagnosis before whatever it was that already happened twice happened again.

For doctors, perhaps the most powerful diagnostic tools available are a phone and a friend. Lagarde immediately thought of Dr. Thomas Duffy. Duffy was one of the smartest doctors she knew, and he was a hematologist. When Lagarde reached him, she quickly outlined the case: a middle-aged woman with two episodes of low blood pressure and a temporary loss of the ability to form blood clots. Did that bring anything to mind?

The phone was quiet for a moment. Then Duffy began to talk through his thought process. The clotting problem did sound like the kind caused by the drug heparin. But there is a type of white blood cell that makes heparin within the body. These cells, known as mast cells, also make another chemical, histamine, which, when released in high doses, can cause low blood pressure—the other mysterious symptom this patient had. Under normal circumstances these mast cells are responsible for allergic reactions like flushing, itching, and hives. (We take antihistamines to block these biological chemicals when we have allergies.) When there is a huge surge of histamine, the body goes into anaphylactic shock—the most severe form of allergic response with a rapid drop in blood pressure, heart palpitations, nausea, and diarrhea, which were all symptoms that this patient exhibited.

"I think it is very likely that this patient has systemic mastocytosis—I cannot think of anything else that would account for this unusual presentation," Duffy offered in his elegant manner of speaking. Systemic mastocytosis is a rare disease in which the body accumulates too many mast cells. When this population of cells is exposed to certain triggers, they dump their huge stores of histamine, and in rare cases heparin, into the bloodstream, causing anaphylactic shock and blood that cannot clot. Certain drugs have been shown to stimulate this reaction in mast cells. This patient had just started taking an antidepressant before coming to the hospital. Was she taking any medications before her previous attack?

Lagarde hurried back to talk with the patient. Yes, the patient said, she had been started on another antidepressant before her last attack, too. Lagarde explained Duffy's theory concerning systemic mastocytosis to the patient. There is no cure for this overgrowth of cells, but patients can manage their symptoms by using antihistamines and avoiding medications that are thought to trigger an attack.

The patient followed up with Duffy, who was able to confirm the diagnosis with a blood test and a bone-marrow biopsy. Since then the patient has carefully avoided antidepressants. But she occasionally feels her heart flutter and stomach turn, symptoms that indicate that her mast cells are acting up for some reason, and she takes her antihistamines, which rapidly neutralize the histamine and reverse the symptoms.

Thinking back, the patient says she has had these symptoms off and on for years. Her heart and stomach would flutter; she would become lightheaded and sometimes a little confused. Her doctors thought these incidents were an overreaction to stress—panic attacks. "I didn't believe it, but when so many people tell you the same thing you can't help but think they are right," she said. "I tried everything—yoga, meditation, exercise." None of it worked. She laughed, then added she now knows that what she really needed was a diagnosis and an antihistamine.

PART III

My Aching Head

Changing Visions

The sixty-three-year-old man in the back seat slept as the car crossed San Francisco's Bay Bridge. His wife sat silent, grateful to her daughter-in-law for driving them to see yet another neurologist and considering how much the man she married thirty-eight years earlier had changed.

It started the year before, with headaches. At first, they were occasional—he noticed them only because he didn't usually get headaches. Then they got fiercer. It was, he told her, as if someone were trying to break through his skull from the inside. They were mostly in the back, mostly on the right, but somehow they managed to hurt his whole head. Lying down helped. Bending over was torture.

Over the next few months, they went from occurring every now and then to being nearly constant. He wasn't a complainer; his wife could tell he had a headache only because he was quieter than usual. But once he ended up on the bathroom floor, his face pressed against the cool tile, the pain so bad he cried.

His regular doctor was worried. It was unusual to start having headaches at his age. She sent him for an MRI. The image was strikingly abnormal. The meninges, the tough tissue that surrounds the brain, usually shows up on an MRI as a crisp thin line. Here it was thick and lumpy. She referred him to a neurologist.

The MRI suggested that something had infiltrated the lining of the brain, the specialist told them. He ordered a second MRI. It looked even worse. Perhaps it was an infection or possibly cancer. But none of the tests of the blood or spinal fluid detected either disease.

Still, the tests weren't normal. Could this be sarcoidosis—a disease characterized by tiny collections of inflammatory cells—or some other inflammatory disorder caused by an immune system attacking its own body? The neurologist referred the patient to a doctor at the University of California, San Francisco, a specialist in these rare inflammatory diseases of the brain.

While making all these doctor visits, the man's wife began to notice subtle changes in her husband. He'd never been a big talker, but these days, when she asked him a question, he'd just grunt or shrug—as if he didn't care. He was clumsier. When he walked or drove, he drifted to the right. At home, he had always been organized. But lately she found the silverware with the plates, his sweaters in her drawers.

Her husband was a painter. Recently he had started working with oil paints, and his bucolic landscapes had taken on darker, more menacing tones. She worried that his perception was changing. On a recent trip to a huge warehouse store, her husband turned to her and exclaimed: "Oh, they changed the layout here." She glanced around; it looked no different to her. The aisles are diagonal now, he said, gesturing off to the right, as if they headed that way. They're not diagonal, she protested. He didn't answer.

The specialist at UCSF didn't think it was sarcoidosis, but he was unsure what it might be. He ordered a third MRI, this one focused on the blood vessels in the brain. These images were also abnormal, but what was causing these abnormalities, or where they came from, wasn't clear. He referred the patient to yet another neurologist, Dr. Wade Smith, a specialist in stroke and other vascular diseases of the brain.

And so the trio found themselves once more headed across the Bay Bridge to UCSF to see this newest specialist. Smith ran through the now-familiar questions about the man's medical history, and then he asked something no one ever had: Do you hear your heartbeat in your ears? The man looked surprised. Yes, he did. Smith then placed his stethoscope over the patient's right eyelid and listened. After a minute, he moved his stethoscope to the area just behind the patient's ear and listened. I can hear what you hear, he said. And I think I know what you have.

When patients can hear their heartbeat in their ears—what's known as pulsatile tinnitus—it's usually caused by turbulent blood flow that makes a noise loud enough for the patient, and sometimes others, to hear. Often this is caused by blockages that narrow and distort the carotid artery, but anything that disturbs the flow of blood in vessels near the ear can do it.

The noise, the headaches, and all the other problems he was having, Smith explained, were caused by abnormal connections between the thick muscular arteries that carry the blood pumped out of the heart and the narrower, more delicate veins.

The effect of these abnormal connections, known as fistulas, is to slow everything down, essentially cutting off blood flow in or out of the region. It's as if a busy interstate is redirected onto a residential street. Everything comes to a standstill. Until the fistula is repaired, the traffic can continue to back up and the congestion spread to a wider and wider area. This backup increases the pressure in the veins, which in turn raises the risk of bleeding.

This type of fistula is quite rare, and while it can be a defect that is present from birth, in adults it is more likely a result of trauma, Smith explained. The man's wife nodded. She told the doctor that she and her husband were in a terrible accident ten years before. Hit by a drunken driver going far too fast, their car was sent tumbling off the freeway and into a tree. Her husband was slammed up against the airbag with enough force to crack his sternum and injure his heart. It was after that accident that the patient started hearing his heartbeat in his head, mostly on the right side. He asked his doctor about it, and she reassured him that it was fairly common. (And it is.) So, he stopped mentioning it. No one had ever asked about it, until now.

Patients with a fistula in the tough outer lining of the brain usually have headaches and sometimes pulsatile tinnitus, the doctor explained. But the man had other symptoms as well: His perception changed, as evidenced by his paintings and his driving. His speech and thinking slowed. These new symptoms suggested that the congestion and poor blood flow had now extended beyond the meninges into the brain itself.

Repairing damaged vasculature is delicate and painstaking work. Dr. Van Halbach, a specialist in this type of procedure at UCSF, introduced a thin catheter into a large vein in the top of the patient's thigh and carefully advanced it, making his way through a highway of vessels until he reached the brain. He injected dye, searching for the hundreds of dysfunctional connections between artery and vein. Once located, they had to be closed off completely, the way you might patch a leaky hose. Only when these abnormal connections were closed would blood flow be fully restored. It took Halbach over eighteen hours to repair the damage that had started a decade earlier in that high-speed crash.

After the surgery, the patient finally began his slow recovery. It took three long years before he started to feel like the man he used to be. His wife agrees. He was talking again. He was driving again. And his newest paintings—bright and colorful once more—prove that the man she married finally came home.

It Started with Sinus Pain

In the early-morning hospital dimness, the man awoke to see a platoon of doctors surrounding his fourteen-year-old daughter's bed. He could see her sitting up, mouth open, a glint of sweat on her cheeks and forehead. He could hear the rapid, ragged breaths, almost as if she had just run a race. She looked over to him. She was scared. And suddenly, so was he.

"We have to take her to the intensive care unit," one of the doctors announced quietly. Equipment there could help her breathe more easily. As the nurses packed up IVs and hooked up portable tanks of oxygen, the girl's father and mother gathered their books and bags. His wife seemed much calmer than he felt. Until that moment, he hadn't believed his daughter was that sick. She was healthy—a star on the soccer team—and a little bit of a drama queen. Sure, she had looked uncomfortable when they took her to see Dr. Suhaib Nashi, her pediatrician, that morning. She was already breathing fast, as if there wasn't quite enough air around her. But even as Dr. Nashi sent her to the Morristown Medical Center in New Jersey, he offered reassurance to the worried parents. And when they admitted her to the hospital, the doctors in the ER said it was just to get on top of this pneumonia by giving her intravenous antibiotics.

But now the look of terror on his daughter's face—and the concern on the doctors' faces—made him see that his daughter was critically ill. He felt tears in his eyes. His wife nudged him. "Don't," she said, glancing toward their daughter.

A few months earlier, his daughter started to get headaches. At first, he wondered if she was just trying to get out of going to school. But Dr. Nashi and the ear, nose, and throat doctors she saw agreed that the headaches were a result of sinus infections. The father had had sinus trouble his whole life, so he knew that kind of pressure and pain. But despite staying home from school sometimes, his daughter could feel well enough in the afternoon to play soccer, and so he wondered whether she was exaggerating.

Her mother didn't have these doubts. She could tell their daughter was in pain. And Dr. Nashi certainly took her pain seriously. When the girl didn't respond to a couple of antibiotics, he sent her to an ENT. Over the course of that spring and summer, the poor girl had been on some half dozen different medications trying to get rid of the headache-inducing sinus infections. It felt as if they were in some doctor's office every other week.

Then, in mid-July, the normally active fourteen-year-old started spending every day lying on the sofa watching TV.

She felt weak and tired, she said. And her head throbbed; her joints ached; her ears hurt. By August, she was asking for help getting up from the couch to go to the bathroom.

And she was worried: It definitely wasn't just a sinus infection, she told her parents. She was sure it was cancer. When she got a strange rash on her elbows—bumpy, red, and not at all itchy—she was convinced it was Lyme disease. When it wasn't, she worried about cancer again.

In the ICU, the doctors put a mask over the girl's face to force air into her lungs. It was awful to see her, terrified, nearly hidden behind all the equipment. But it seemed to help.

Dr. Simona Nativ, a pediatric rheumatologist from the nearby Goryeb Children's Hospital, saw the family late the next afternoon. She had heard about the chronic sinus infections that seemed untouched by antibiotics, and she had an idea of what might be going on. Did the girl have a rash on her elbows? The mother, amazed, asked how she knew. The doctor said it was a symptom seen in one of the diseases she had in mind. She examined the daughter, starting with the rash. Probably, the rheumatologist said, this was not an infection. A biopsy of the rash would be helpful; so would some additional blood tests.

The possibility of an answer had given them hope, but it was lost late that night. The girl, still short of breath and with a worsening cough, was surprised to see that the tissue she used to cover her cough was bright red. We've got blood, announced a nurse, holding out the tissue, and the atmosphere in the room shifted. This wasn't just pneumonia after all.

They needed to see what was going on in her lungs, yet another doctor explained. After giving the girl a little sedation, he snaked a camera through her mouth and into her airways. There was no sign of infection. Instead, her lungs were filled with blood and clots. There were a handful of causes of this kind of major bleeding. They were all unusual: a few infections, some tumors or, if she were a baby, some aspirated object lodged in her lungs.

Because of the blood and her trouble breathing, the young woman was put on a ventilator. The race was on to stop the bleeding and figure out the cause. If she kept bleeding, she would die.

Samples of blood and lung fluid were sent to the lab in search of a diagnosis. But it was the blood and the tissue biopsy from the rash that provided the answer. The girl had something called granulomatosis with polyangiitis, or GPA. It is an autoimmune disease—an illness caused by her own antibodies, the foot soldiers of the immune system, which mistakenly attacked the blood vessels in her lungs. It had also injured the tissues of her airways and sinuses, producing those initial headaches. It even caused the rash on her elbows.

It's not known what causes this unusual disease, and this girl was an unlikely target. GPA is most commonly seen in adults over age sixty. It can be devastating: Untreated GPA has a mortality rate of roughly 80 percent at one year. Treatment involves powerful drugs that target the cells making the antibodies: high-dose steroids plus one of two fierce immune-suppressing medications borrowed from cancer chemotherapy. Eliminating the deviant cells seems to allow the immune system to reset. And when the drugs are stopped, the self-directed foot soldiers are usually gone.

They often come back, however. Many patients will get doses of the immune-suppressing medication once or twice a year to prevent a recurrence. The high-dose steroids can be started as soon as the diagnosis is made. But the chemotherapeutic agent is so effective at suppressing the immune system that before it can be administered, her doctors had to be certain there was no hidden infection that could come roaring to life.

When no viruses or bacteria were found, the girl was given a medicine called rituximab. Within days, she started to improve. But it took nearly two weeks for her lungs to clear enough to allow her to breathe on her own.

Her recovery was delayed by complications from the disease as well as the treatment. Blood clotted in her arms and legs and traveled to her lungs. The steroids made her so weak that once she got off the ventilator, she could do no more than just breathe. She couldn't eat, talk, or even hold her phone. A couple of weeks later, when she could finally walk with help, her parents took her home. It was weeks before she could make it to the bathroom alone, months before she could go back to school part time. She worked hard to catch up on her schoolwork.

She never went back to playing soccer. She just didn't have the endurance. For years afterward she had nightmares that she was back in the hospital, too weak to move and consumed by the fear that she would never get better. But she did. And after graduating from high school she started college. Her goal is to be a nurse. Although she's a little nervous about returning to a hospital, it was the nurses at her bedside who made her feel better, even while she was still quite sick. And she hopes to someday provide the same comfort and care for children, who, like her, need so much of it to get through a terrible illness.

The Elephant Trainer Gets a Headache

It was a cool fall day, but the sun seemed extremely bright as the young man helped guide nine circus elephants to their new pens. Even though the man was wearing sunglasses, the morning sun reflecting off the metal equipment felt like a knife cutting into his right eye. His head throbbed behind the eye, and an occasional tear rolled down his cheek. When the animals were finally secured, he returned to his trailer. "Okay, I do need a doctor," he said to his girlfriend. His hand was cupped over the side of his face. "Right now."

It was the worst headache of his life, the twenty-five-year-old patient told the doctor in the emergency room of Highland Hospital in Rochester, New York. It started five days earlier when the circus was in Connecticut. At first it wasn't a big deal. He would take a couple of aspirin, and it would disappear. But when the medicine wore off, the headache was still there. In fact, each time it seemed just a little worse. That morning, when he got out of bed, the pain was unbearable. He took aspirin, Advil, Tylenol. Nothing put a dent in it. The pain was sharp and on the right. It felt as if someone were slamming a door inside his head. He'd had the occasional headache but never anything like this.

He didn't smoke, rarely drank, and took no medications. He had no recent head trauma, though he was head-butted by a zebra a few years before. That hurt—it broke his glasses—but not this much. His mother had migraines, and perhaps that's what this was. Maybe, the doctor said, though a week was a long time for a migraine.

For doctors, a description of a headache as the worst is a red flag. We worry about headaches described as the first (for someone who doesn't have headaches) or the worst (for someone who does) or those that are "cursed" by the presence of other symptoms like weakness or confusion. He didn't have other symptoms, but the doctor was concerned because he called it the worst.

The doctor ordered a painkiller and blood tests to look for signs of infection or inflammation. She also ordered a CT scan of the head to look for a tumor or evidence of blood. The blood tests were normal. The CT was not.

Within the brain, there are compartments where spinal fluid is made. The fluid then circulates around the brain and spinal cord and is reabsorbed. Two of these compartments, known as the lateral ventricles, are usually mirror images of each other. But in this patient, the ventricle on the right, where his headache was located, was much larger than the one on the left. That suggested there might be a blockage in the circulation of the spinal fluid on the right side, which was causing pressure to build. That could certainly cause a headache—and permanent damage if not addressed quickly.

Even before the ER doctor saw the CT scan, she called neurology for help in figuring out this patient's terrible headache. The neurology resident examined the patient and his CT scan, but it wasn't clear to him how the pieces fit together. If the asymmetry were caused by an obstruction, the patient should have symptoms associated with increased brain pressure—like nausea—but he didn't. The resident knew that he didn't have enough data to make a diagnosis. Watching the patient over time would give him more. If there was a blockage in his brain, he should begin to feel nauseated and weak. If he didn't, it was very unlikely that the asymmetry reflected a blockage. The patient was admitted to the hospital, where nurses were to examine him every four hours to look for any change.

Overnight the headache became worse, despite the use of several powerful painkillers. By morning the patient was exhausted from the pain and nearly incoherent from the narcotics. He never, however, developed symptoms of increased pressure in his brain. The neurologist speculated that this was a migraine and recommended he go home and follow up as an outpatient.

The neurosurgeons weren't so sure there wasn't an obstruction. The patient's worsening pain was worrisome. They recommended an MRI. If there was a change in the size of the ventricle, when compared with the CT, they could drill a small hole into his skull and relieve the pressure.

Dr. Bilal Ahmed, the internist taking over the patient's care that morning, first heard about the new patient from his team of residents outside the patient's door. They told him that he was a young circus worker who had been hit in the head by a zebra, had an abnormal CT, and was probably going to surgery later in the day.

As they stood there, a nurse hurried out of the patient's room. "He's got a rash," she told the doctors. The team went into the room, and Dr. Ahmed glanced at the patient now hidden beneath a pile of blankets. He introduced himself to the patient's girlfriend. As she started to speak, Dr. Ahmed held a finger to his lips. "Don't say anything," he told her. "I want to see for myself."

"May I look?" he asked the young man. A matted head of dark curls slowly emerged from beneath the mound of blankets. The patient sat up slowly, blinking in the dim light. His right eyelid was swollen and drooped drunkenly over the pupil so that only the lower ridge of the greenish brown iris was visible. The right side of his forehead was red, as if he had a sunburn on that half of his face. And there was a sprinkling of bumps over his eye and forehead.

Was this zoster? Dr. Ahmed wondered out loud. He touched the reddened skin around the lesions. The young man winced. That part of his forehead had been intensely sensitive ever since this headache started.

Herpes zoster—or shingles—is the reemergence of the herpes virus that causes chickenpox. The word "shingles" comes from the Latin *cingulum*, which means "belt" or "girdle"; the rash of herpes zoster often appears in a band, usually on the trunk or chest. When a chickenpox infection resolves, the virus takes refuge in branches of the nerves just outside the spinal cord, where it usually resides for decades. Sometimes the virus reemerges, but the reasons are unclear. Most of these outbreaks are painful but not dangerous—except when the virus emerges in the nerves near the eyes.

Dr. Ahmed called the neurosurgeon. Was there a link between this patient's shingles and the asymmetric ventricles? No, he was told. If this guy has shingles—and it sounded as if he did—then the asymmetry was probably something he was born with. The MRI, done later that day, confirmed that there was no obstruction. In the meantime, the patient was started on an antiviral medication. Despite the treatment, his vision began to blur. The bumps on his face, which led to the diagnosis, had spread to his eye as well. Two years later, his vision is still impaired on that side.

In this case, as in so many, time is a powerful and frequently undervalued diagnostic tool. The rash appeared days after the symptoms began; that is common in zoster. But without the telltale rash, there was only the pain and the abnormal CT, and that led his doctors to worry that his pain was the result of pressure building up in his brain. A truism in medicine is that when we hear hoof beats we should think of ordinary horses as the cause rather than the rare zebra. In this case, time revealed that what looked likely to be a zebra—an obstruction on the right side of the brain—was actually the everyday horse of herpes zoster.

A Sea of Gray

"I can't see a doggone thing. It started off as a headache, and now I can't see." The middle-aged man's face was flushed and shiny beneath a mop of prematurely white hair. His clear blue eyes were shaded by a brow pulled together by worry. "I'd rather cut off my leg than lose my sight," he told the slender, dark-haired doctor at his bedside. Three days ago he was at work at a local animal hospital when his head began to pound. "It was like there was someone inside my head trying to get out." He made it to the end of the day, then went home and straight to bed.

The headache was still there the next morning when he got up. He took his coffee and the Sunday newspaper to the living room and—as was his habit—turned to the obituaries. The page was a sea of gray. He couldn't read it. He couldn't even make out the headlines.

He went to see his doctor, who sent him to an ophthalmologist. The ophthalmologist sent him to the emergency room. Hearing that his patient was being admitted to a hospital, the primary care doctor promptly called Dr. Lydia Barakat, an infectious disease specialist at Waterbury Hospital in Waterbury, Connecticut. The fifty-eight-year-old man had a high fever, he reported, and the thick optic nerves that connect the eyes to the brain were visibly swollen from some kind of increased pressure inside the skull.

Barakat was worried. Infections in the brain carry a high risk of death and disability. "If you lose a nerve cell, it doesn't come back," she explained to me. Infections involving the brain were those most likely to keep her up at night. These were the infections where you couldn't afford to miss anything. "If you're not just a little scared when you see these patients, then you are either arrogant, indifferent, or just plain ignorant."

The patient had diabetes and high blood pressure, but he took his medicines regularly, and they hadn't caused him trouble for years. He didn't smoke and didn't drink. He had been married for thirty-eight years.

On exam, his skin was warm and damp from a fever of 101, and he still had difficulty seeing. The rest of his exam was unremarkable. Dr. Barakat reviewed the studies that had already started to fatten up the patient's chart. He'd had routine blood work, a CT scan, an MRI. All were normal. The spinal tap, however, was not. When the doctor first placed the needle into the sac surrounding the spinal cord, pale liquid, which normally comes out one drip at a time, gushed out, confirming that the patient had elevated pressure in his central nervous system. And the normally cell-free fluid contained a handful of white blood cells.

It was clear that the patient had meningitis, an infection in the tough tissue that envelops the brain. It was also clear he probably didn't have the deadliest type. The most aggressive types of meningitis can kill within hours, and this man had been sick for several days.

The most common causes of meningitis are viruses. These are usually less severe infections and will often resolve without treatment. But it would take days to know for certain if this was viral. And there were other, more unusual possibilities where antibiotics would be essential. Could this be Lyme disease? Although the tick-borne infection was most common in warmer months, and there was still snow on the hard, frozen ground, this was Connecticut, and the disease was endemic. So Lyme had to be considered. Ditto the mosquito-borne West Nile virus. This kind of vision loss wasn't classic for either of those infections, but both agents cause fever and frequently invade the central nervous system.

Or did the patient have a more colorful personal life than he was allowing? Syphilis could cause vision loss. Usually the loss of sight came on gradually, and a fever would be rare.

Still, if he did have syphilis, antibiotics could save his vision. A recent infection with HIV could also cause a meningitis-like illness. The patient didn't have any HIV risk factors, but when Dr. Barakat suggested he be tested—just to be certain—the patient refused. Was he hiding something?

Finally, the patient told her he was bitten by a cat two weeks earlier. Cat scratch fever rarely causes meningitis, and he had none of the swollen, tender lymph nodes usually seen in this kind of infection. Still, he would need to be tested for each of these possibilities. And Dr. Barakat would ask him again about the HIV test. In the meantime, he was already on high doses of two antibiotics that would cover the other bacterial causes of meningitis on her list.

By the next morning, the patient's fever had come down

and the headache, while still present, had improved. His vision, however, was just as bad. The patient's first questions, when Dr. Barakat visited him, were about his sight. Would he be able to read again? Would he be able to work again? She tried to be optimistic but explained that it's hard to predict what will happen without a diagnosis. She asked him again about the HIV test. "I love my wife and I always have," the patient told her simply. He had never been unfaithful. And he was certain the same was true for his wife. An HIV test was unnecessary. Dr. Barakat nodded. It was hard to believe he would withhold this kind of information when the stakes were so high.

The results of the studies trickled in over the next few days. It wasn't Lyme. It wasn't syphilis or West Nile. But the test for *Bartonella henselae*, the bacteria that usually causes cat scratch fever, was interesting. Since *Bartonella* is difficult to grow in a petri dish (which is how most bacterial infections are identified), the testing looked instead for the presence of antibodies to the bacteria. If he had ever been exposed, the test would pick up some antibodies, but if he had the infection now, the number of antibodies should be high. This patient had antibodies, but they were in the normal range. Was this the earliest stages of infection, when the antibodies are only beginning to be produced? Or were these antibodies left over from an earlier exposure?

Cat scratch disease, as it is now known, is usually characterized by swelling at the site of the bite or scratch, with fever and enlarged, tender lymph nodes. It's generally seen in children and transmitted by kittens. In this case, both the cat and the patient were in the wrong age group. The only way to know for certain was to retest him in a few weeks. If cat scratch disease was the culprit, the measured antibodies should be much, much higher. Dr. Barakat stopped all the antibiotics except the one for *Bartonella.* This was likely a virus, but cat scratch disease was still a contender.

The patient's headache disappeared after three or four days, but his vision remained poor. He was started on prednisone to reduce the swelling in his optic nerves and sent home to finish up his antibiotics by mouth. Over the next two weeks, the patient's sight improved, and he went back to work at the animal hospital. His diagnosis was still not clear. "They told me I had a meningitis, and they didn't know what caused it," he told me.

The answer finally came a month later when a second blood test revealed sky-high levels of antibodies. It was cat scratch disease. "It's funny," the patient said. "I had to get well to get an answer."

A year after his stay in this small community hospital, he was still having a little trouble with his color vision (he couldn't tell white from yellow). The way the patient tells it, his illness simply confirmed his longstanding position in the great dog-versus-cat debate. "I've always hated cats," he said. "They give me the creeps." He usually works only with dogs. But on that day, he told me, he was the only extra pair of hands available, and so the vet asked him to hold a cat that needed a shot. The needle-sharp teeth penetrated the skin between his thumb and forefinger. "It hurt like the blazes, but my hand never got red, never got infected." He paused, then smiled broadly. "And I thought 'Cat Scratch Fever' was just a song. Who knew it could just about kill you?"

Everybody Lies

The boy in the speeding ambulance lay with his eyes closed for most of the trip to Beaumont Hospital in Royal Oak, Michigan. When the pounding in his head subsided enough for him to open his eyes, he noticed the concerned look on the paramedic's face. What's she so worried about? he recalls wondering. That was the last thing he remembered until after his surgery.

The headaches started the weekend after Thanksgiving, just two weeks earlier, when the fifteen-year-old woke in agony as pressure encircled his skull. He had had headaches in the past, usually after a football injury, but they were nothing compared to this. The next two weeks were a blur. He wasn't hungry or thirsty. He stopped going downstairs for meals. Opening his eyes made the pain so unbearable that it was easier just to stay in bed and keep them closed.

His mother took him to his pediatrician. Any fever? the doctor asked. Any nausea or vomiting? These are symptoms of infection or increased pressure in the skull. Nothing but the pain and the terrible sensitivity to light, the boy reported.

Migraine, the pediatrician reassured them. He's at an age when they often start. They can last for days—give it time. Time didn't help, so the mother took him to the emergency room. Migraine, they were told again, or maybe a viral meningitis. The boy was given a stronger painkiller and sent home.

Days passed, and nothing changed, so they returned to the doctor. Still no fever, nausea, or vomiting. It was probably a viral meningitis, they were told. He will get better. The mother asked about a CT scan of his head. It won't help, the pediatrician said to them. Headache is one of the most common complaints seen in primary care. Ninety-nine times out of a hundred the cause won't be visible even with the best imaging.

The parents became increasingly distraught as their active and hard-working son, an A student and high-school quarterback and baseball player, spent days in bed, hardly able to eat. His skin took on a grayish color. He looked as if he was dying. Finally, his mother decided to take him back to the hospital, determined to get a CT scan of his head.

In the ER, the doctors asked the same questions: Any fever? Any nausea or vomiting? But this time the boy's mother had different answers. Yes, he had been having fever for days. Yes, he was so nauseated he could hardly keep anything down. Lying to the doctors didn't feel right, but the mother thought that symptoms that fit their expectations would get action faster than her own motherly observations. Perhaps her son would have gotten a CT scan without her "fabrication" (as she calls it), but she didn't want to take that chance.

Soon after the scan, the ER doctor approached her. The CT scan showed something in his brain. The doctor wasn't sure what it was, but it was big and ugly. The boy had to be transferred to the nearby pediatric hospital. The paramedics bundled him into an ambulance, and his mother followed in her car as they headed to Beaumont Hospital.

At Beaumont, an MRI showed the boy had an abscess and needed emergency surgery to reduce the pressure in his brain. Hours later, he awoke in the intensive care unit confused and scared, but finally free from pain.

The next day, Dr. Bishara Freij, a specialist in pediatric infectious diseases, came to see the boy. This type of abscess is usually a result of an infection in the ears or sinuses or teeth that extends into the brain, but this patient showed no sign of any other infection. The boy's exam was completely normal, except that the inside of his nose was caked with blood, evidence of a nosebleed.

The blood tests were likewise unremarkable—no hint of an infection. The brain scans confirmed the presence of the abscess, but nothing more. The chest X-ray was uninformative.

When Dr. David Bloom, a senior pediatric radiologist, reviewed the film, he did not think the boy's chest X-ray looked normal. There was a dim but discrete area of lightness where the lung should be mostly dark. What was that?

A favorite radiology teacher of his once told Bloom that being smart was good, but having old films was even better. So Bloom looked at the patient's older X-rays. He found the same abnormality. Based on its location and appearance, he thought it probably represented a pulmonary arteriovenous malformation (AVM), an aberrant connection between the arteries and veins. Normally, blood flows from the arteries to the capillaries, a system of tiny vessels where oxygen is delivered to bodily tissues and waste products are collected to be carried through the veins to filtering organs. When the capillaries are bypassed, as they are in AVMs, the waste that should be routed to the venous system for disposal can spread throughout the body through the arteries. If it lodges in the brain, it can cause a stroke or, as in the case of this boy, an infection.

The usual treatment for AVMs is to place coils, tiny pieces of wire often no wider than a human hair, inside the abnormal vessels. Once there, they cause clots to form, blocking the flow of blood. When Freij heard that the boy had an AVM in his lungs, he immediately knew the diagnosis. Most patients with pulmonary AVMs have a disorder known as hereditary hemorrhagic telangiectasia, or HHT. Those with HHT have thin, weak, easily damaged blood vessels, which tend to dilate and rupture. When this happens in the superficial layers of the skin, the bulge caused by the dilated vessel can be seen in the form of tiny red spots that disappear briefly when pressure is applied. These spots, called telangiectasias, are prone to bleeding, especially when they form on the more delicate mucus membranes of the lips, mouth, nose, and gastrointestinal tract. Indeed, frequent nosebleeds are a hallmark of the disease. The bleeding can be profuse, and sometimes necessitates blood transfusions. Patients with the disorder also have a propensity to develop AVMs in the lungs, liver, and brain.

Until the end of the twentieth century, a diagnosis of HHT was based exclusively on clinical and radiographic findings. Frequent bleeding from the nose or gastrointestinal tract along with a family history of the disease or the presence of telangiectasias or AVMs were the diagnostic criteria.

Any three of the four were enough to define a case of HHT. In 1994, the first gene associated with the disorder was identified. Since then, more than six hundred genetic mutations have been linked to it.

After weeks of treatment, the patient finally went home. The most significant aftereffect was a permanent loss of vision on his right side. He spent weeks working with an occupational therapist learning how to compensate for this loss.

He returned to school with just two weeks left in the semester. Despite having to make up twenty-one tests, including midterms and finals, he ended up with straight As. He was a little disappointed, he told me, that one was an A–. His mother knew he was fully recovered the next summer when he pitched in a championship baseball game. Seeing him on the mound, looking as he always had, brought tears to her eyes.

Several years later, the young man was working on a PhD in physiology and planning to pursue medical research. His goal: to find a cure for HHT.

The Worst Ice Cream Headache, Without the Ice Cream

It was the tuna salad sandwich that did it, the patient—a friend—told me. He was eating the sandwich when an excruciating pain tore through his throat, his jaw, his ear. He dropped to the floor and grabbed his face. He rubbed, he massaged, he flexed his jaw. Nothing he did relieved the knife-stroke of pain that consumed the entire right side of his face. After what seemed an eternity but was probably only a few minutes, the pain began to ebb. After another ten to fifteen minutes, it receded to the persistent ache that had been his constant companion for the past two weeks. Do you know what this is? he asked me. I didn't.

It all started out as a sore throat, he told me. Maybe he was coming down with a virus, was his first thought. Then, a couple of days later, his teeth began to hurt. Not all of them. Just the last two molars in the back, on the right. Eating or drinking anything—hot or cold—would set off a pain like the one caused by eating ice cream too fast, only much, much worse. It was by far the most severe pain he'd had since he passed a kidney stone—an event he could still vividly recall after more than twenty years.

The gnawing ache sent him to the mirror to look for a possible source. He poked and prodded his teeth. Nothing. He'd gone to the dentist for his annual checkup just a couple of weeks earlier and received a clean bill of health. Then why did his teeth hurt so much? Over the next several days, the pain spread to the entire right half of his face. It was as if he had a sore throat, a toothache, and an earache all at the same time. Constantly. Every now and then, especially when he ate or drank, paroxysms of pain would shoot from throat to ear.

He was forty-eight, active, fit. It had been so long since he'd been sick that he no longer even had a doctor. So he asked his father, a neurosurgeon: Do you think it could be my sinuses? Should I take some antibiotics? They can't hurt, his father told him. Maybe not, but they didn't help much, either. If anything, the ache worsened and the occasional shots of ice-cream pain seemed to get more frequent and more intense.

He went to a local clinic. The doctor on duty looked in his ears, his nose, his mouth. Nothing. No fever. No redness. No enlarged glands. He did a rapid strep test. Normal. It was probably some kind of virus, the doctor told him. Fluids, aspirin. It should get better on its own.

But it didn't. And then he ate the sandwich that brought him to his knees. There must be someone who could tell him what this was. When I had no answer, he went to the yellow pages and picked out an ear, nose, and throat doctor in nearby Falmouth, Massachusetts—Dr. Douglas Mann. Yes, they told him, the doctor could squeeze him in that afternoon.

The doctor was young-looking, with a pleasant, businesslike approach. "I understand you're having some trouble with your throat," he said. The patient explained the events of the past couple of weeks. "By the time he got to the end of his story, I knew what he had," Mann told me. Still, he needed to be sure he was right. The list of problems that can cause throat pain is long and varied, but this patient complained of pain on only one side of his throat. That shortened the list significantly. And he described a pain that traveled from the throat to the ear, limiting the possibilities to a mere handful. Mann had to be certain it wasn't cancer. The patient had a history of occasional tobacco and alcohol use, a combination linked to a higher risk of head and neck cancers. An aphthous ulcer, also known as a canker sore, can cause severe unilateral throat pain that often radiates to the ear with swallowing. So can a tonsillar abscess. All of these could be ruled out with a thorough physical exam.

Mann looked the patient over as he listened to his story. He looked healthy—slender and tanned. This made an invasive cancer much less likely. Mann meticulously examined the patient's ears, nose, mouth. Like the doctor before him, he found nothing. Since the patient's pain had started in his throat, the doctor would need to examine it thoroughly as well. Mann sprayed a topical anesthetic into the patient's nose, then pulled out an odd-looking instrument with a long, slender black tube on one end and an eyepiece on the other. He snaked the flexible endoscope into the patient's nose, through the oropharynx, and then down his throat. Mann stared through the eyepiece on the body of the machine as it traveled through the dark alleyways, peeking into the dim nooks and crannies. Nothing.

Mann smiled, confident at last that his initial impression was correct. "Didn't you tell me that your father was a neurosurgeon?" The patient nodded. "He'll get a real kick out of this diagnosis." What he had, the doctor explained, was a condition known as trigeminal neuralgia, or "tic douloureux"—painful tic. In this disorder, a facial nerve—known as the trigeminal—misfires, overreacting to even insignificant stimulation. A simple act like eating or swallowing, or even a light touch, can trigger the spasms of excruciating pain the patient described. The name derives from the characteristic grimace often made in response to the pain, but it's something of a misnomer, since a tic is an involuntary and uncontrollable muscle twitch or spasm. Still, it's an apt description.

Tic douloureux primarily affects those over fifty, women more commonly than men. While the disease was first described more than a thousand years ago, its cause remained a mystery until very recently. We now know that this disorder usually occurs when a blood vessel compresses one of the nerves providing sensation to the face as the nerve leaves the brain to travel to the skin of the forehead, cheek, mouth, or throat. This pressure erodes the nerve's protective covering so that it fires off with little or no provocation. Why this happens is still unclear. There are other diseases that can mimic this condition, the doctor added. Rarely, multiple sclerosis causes a similar syndrome. Even more uncommon, a brain tumor impinging on the nerve can produce the same symptoms. Given these possibilities, the patient would need an MRI.

With this condition, however, the primary treatment goal is pain control. Sometimes the syndrome will resolve on its own, but it can last for years. Mann gave the patient a powerful painkiller for immediate relief and started him on an anticonvulsant medication that has been shown to be effective in treating nerve pain. Surgery can relieve the pressure on the nerve as well, but like all such operations, it's a tricky business and to be avoided if at all possible.

The next time I spoke to my friend, he was feeling much better. "Today was the first day I've woken up without pain," he reported, barely a week after starting the medication. He had called his father to tell him his diagnosis. "Dad told me about all the lovely neurosurgical procedures I might be in for if the medication route doesn't work." He laughed. One involves threading a needle through the face and destroying the nerve, he told me. In another, the surgeon drills into the skull and moves the artery away from the nerve. "So I told him to scrub up. Whether he's retired or not, I'll be expecting him to do the cutting if I ever need any."

An Icepick to the Head

A fifty-four-year-old woman sat quietly as the chiropractor worked her neck the way she had so many times before. Suddenly a pulsing, whooshing noise thundered in the woman's left ear. It reminded her of the sound of her heart when she had an echocardiogram years earlier—but this was much, much louder. The chiropractor stopped immediately, but the noise did not.

It pounded away at her ear—rhythmic, loud, and unrelenting. The noise was always present but seemed to get worse when she lay down or turned her head a certain way. Usually it was just annoying, but sometimes it grew so loud she had trouble hearing other people speak. Still it seemed manageable, at least at first.

A few weeks later, she suddenly felt as if she had been hit on the left side of her head with a brick. The pain was blinding. She had to leave work to lie in a darkened room. It was probably a migraine, her doctor told her, giving her a prescription for a drug called Zomig and ordering an MRI of her brain. Neither the medicine nor the scans were helpful. The headache lasted two more days. After that, the occasional migraine would come and go; the whooshing noise remained constant.

Her doctor sent her to a neurologist, who ordered a different MRI, this one of the brain's arteries, to look for any damage from the chiropractor's maneuver. When the test came back normal, he diagnosed tinnitus (the perception of sound in the head or ears—from the Latin word *tinnire*, meaning "to ring") and migraine headaches.

An ear, nose, and throat specialist gave her a hearing test, which revealed mild hearing loss. The ENT then ordered an MRI of the venous system of the brain. It, too, was normal.

A second ENT reviewed the studies and also suggested a diagnosis of tinnitus and mild hearing loss; he added eustachian-tube dysfunction. None of the specialists had any advice on how to get rid of the noise or the headaches.

A few months later, the patient read about a disease called fibromuscular dysplasia (FMD), an unusual disorder in which the walls of arteries narrow and limit blood flow to essential organs—usually the kidneys or brain. Could she have FMD? She made an appointment with Dr. Jeffrey Olin, the director of vascular medicine at Mount Sinai Medical Center, in New York, who was quoted in the article. Olin arranged for detailed imaging of the carotid arteries, the vessels that bring blood from the heart to the brain.

As the technician was injecting contrast dye into her vein, the patient suddenly felt an icepick of pain shoot from the top of her skull past her ears down to her clavicle. It was all she could do not to scream. It was worse even than the migraines she continued to get. Olin was immediately concerned. Could the slightly increased pressure caused by the injection of contrast have torn a fragile segment of the patient's carotid artery? He saw no evidence of a tear, but the images showed that her carotids were not normal. They twisted and turned on their way from the heart to the brain. And on the left, where the pain, headache, and noise originated, the artery made a 360-degree loop.

Olin knew that this kind of abnormality could cause tinnitus and migraines. But he could not explain the pain that occasionally knifed through her face to her neck. A second neurologist thought the pain might be from injured neck muscles. He prescribed a muscle relaxant. It didn't help.

A vascular surgeon gave a diagnosis of temporal arteritis—inflammation of the arteries of the head, eyes, and face, which can cause blindness and strokes. She was started on a high dose of steroids and referred to a rheumatologist. The rheumatologist ordered an ultrasound and then a biopsy of the artery. Both were normal and the patient was weaned off the steroids. The pain and whooshing remained.

Last August, the patient came to my office in Waterbury, Connecticut. Could I figure out the cause of the pain she'd had for the past two and a half years? she asked. She used to skydive, hike, and climb mountains. Now simply going up a flight of stairs could cause excruciating pain in her head.

In medicine, it is essential to rule out diseases that can kill, and then move on to those that may only make you wish you were dead. The patient didn't have a tumor. And none of the scans identified a tear. Still, given her story, I thought it was pretty likely that the chiropractic maneuver had caused some minor trauma, and that the injection of contrast had as well. Could these injuries be contributing to the pain? When I examined her, there was a tenderness in her neck over her left carotid artery. Could she have carotidynia (from Greek, meaning painful carotid), an unusual but well-described condition caused by inflammation of the tissues of the carotid artery? The cause is unknown, but it is most frequently seen in patients with migraines and can usually be treated with the medications used to prevent those headaches. I started her on one and asked her to come back in a few weeks. She was hopeful, and so was I.

Meanwhile, I was busy studying for a test that board-certified internists must take every ten years. As I was reading, I came across a reference to an unusual disease with a Victorian-sounding name: hemicrania continua. I didn't remember much about it and so went to Google to read more. The first site I clicked on—a patient's description of her headache—immediately reminded me of this patient's pain. I moved on to the medical literature, fascinated.

Hemicrania continua is a type of daily headache characterized by persistent pain on one side of the head punctuated by episodes of sharp pain. The painful flare-ups are often accompanied by other symptoms, including watery eyes, runny nose, eyelid swelling, or constriction of the pupils. Remarkably, most patients with this type of headache get better when treated with an inexpensive medication that has been around for years: indomethacin.

My patient hadn't told me about any eye symptoms. Still, could she have this rare headache? I called her. The headache and the whooshing were still there. The medications I prescribed had not worked. Did she have any eye-watering or eyelid swelling when the pain in her head was most intense? I held my breath: if she had these symptoms, then hemicrania continua was a real possibility.

Yes, she said. Sometimes she felt as if she had a cold just in her left eye. Was anything unusual about the pupil in that eye? There was: when the pain was most severe, she noticed that her left pupil would constrict. The symptoms were so mild that she never thought to mention them, and no one had asked her about them.

Now I was really excited. I told her what I'd found and started her on a two-week course of indomethacin. Several weeks later, I called my patient. How was she feeling? She laughed at my question. A few days after she started the indomethacin, she told me, the headache disappeared. Just like that. But she couldn't talk now. She was heading out with some friends for a hike. She was working hard to get back into shape and back to her old level of activity. She'd be in to see me in a few weeks.

PART IV

I Can't Breathe

A Deadly Itch

"I can't breathe," the woman growled in a husky whisper. Her sister looked anxiously at the clerk behind the registration desk at the University of Iowa hospital emergency room. The woman swayed, her breath was rapid and coarse. Her chest heaved. She pulled at the neck of her sweatshirt—suddenly it was too tight. She pulled it over her head and dropped it to the floor. She was naked beneath the top; she had been in bed when this attack came on.

The fifty-four-year-old woman dropped into a wheelchair her sister found and was whisked into the heart of the ER. The rest was a blur of concerned faces, needles, and medical data. Her blood pressure was dangerously low; her heart was racing. She was given epinephrine and steroids, but it was hours before she could explain what happened that night.

She was staying at her mother's house in rural Iowa, she told the doctors. Just as she was going to bed, she felt a sudden tingling in the palms of her hands. She recognized the sensation immediately: Twice in the past eight years she had felt the same strange itch on her hands and sometimes her feet. Each time it was followed by a terrifying sense of her throat closing.

She drove herself to her sister's house, several miles away, and her sister drove her the rest of the way to the hospital. She opened the car window to let in the frigid winter night air. She struggled to breathe. Black spots swam before her eyes but she willed herself not to pass out.

She had had this kind of allergic reaction twice before but never as severely. She knew from her own research that this was anaphylactic shock—a potentially deadly allergic reaction. After she got the medications, the symptoms resolved. She stayed in the hospital overnight, and when it was clear that the episode was over, she went back to her mother's house. She made an appointment to see a local allergy specialist right away.

The specialist spent nearly two hours going over everything the woman had been exposed to—food, plants, toxins, anything that might have triggered this nearly fatal allergic reaction. There were no new exposures that day; everything she ate or touched was something she had been in contact with many times before and after this latest attack. The most common cause of severe allergic reactions in adults is food, but the allergist couldn't identify any likely suspects. He was mystified. He asked her to share her diagnosis if she ever got one.

For months after returning to her home on Long Island, New York, the woman was anxious about everything she ate, and she worried every night when she went to bed. She always kept a bottle of Benadryl and an EpiPen with her, but still she was terrified about what might happen if she was too far from a hospital the next time.

When her next attack happened—just ten months later—

she was already in the Brookhaven Memorial Hospital, in East Patchogue, New York. She was being treated with antibiotics for a devastating case of gastroenteritis due to salmonella. Her first meal, after days of nothing but clear liquids, was beef brisket with potatoes and carrots. It smelled good, but she had no appetite. Still, she made herself eat a few bites, knowing it was her first step toward going home.

A couple of hours later, she felt a strange itch on the top of her head. She scratched reflexively. Then, recognition hit her like a slap: Not now, she thought. She grabbed the IV pole, still dripping fluids into her system, and ran out to the hallway. "I need a nurse," she shouted. Her heart was pounding and she knew what was coming next. Hospital staffers in scrubs descended on her. Was she having a panic attack? No, an allergy attack, she told them.

They helped her back into bed, and gave her oxygen, Benadryl, and steroids. What happened? someone asked. She told the whole story, plus something that she now realized—every one of her attacks seemed to come a few hours after she ate beef. She didn't go through this every time she ate hamburger or steak; meat was a regular and much-loved part of her diet. But she was pretty sure that she'd had steak—or beef brisket, this time—before each episode.

Her doctors were dubious. New food allergies—especially severe ones like hers—are uncommon in adults. This was much more likely to be an allergic reaction to one of the antibiotics they were giving her. The patient found that theory hard to swallow. That might explain this episode, but what about the earlier ones? She hadn't been on antibiotics then. The doctors had no answer.

• • •

A nurse had a different theory about what happened, one the patient had heard before but never believed. There was some kind of tick, the nurse told her, whose bite could make you allergic to meat. She didn't know much about it. But, the nurse suggested, she should check it out.

The woman had had tick bites before—who on Long Island hasn't? But was it really possible for a bite to produce this crazy reaction? Indeed it was, she discovered when she got home and began doing some research. The bite of the lone star tick—named for a white star-shaped spot on the arachnid's back—could cause an allergic reaction to mammalian meat. The trigger was a sugar, identified as galactose-α-1,3-galactose, and more casually known as alpha-gal, a carbohydrate found in the flesh of all non-primate mammals.

How the tick bite triggers this allergy is not yet known. The link between the tick—whose range extends from southern Florida to Maine and as far west as Iowa—and the resulting alpha-gal allergy was first described in 2009 by Thomas Platt-Mills, a professor at the University of Virginia, who himself developed the disorder. Unlike most food allergies, in which symptoms occur within minutes of consuming the allergen, the alpha-gal allergy is delayed. The symptoms—ranging from a rash to nausea to shortness of breath and anaphylaxis—can appear four to six hours after a meal containing meat. Stranger still, the reaction doesn't occur after every exposure.

The diagnosis of mammalian meat allergy (MMA) can be confirmed with a blood test that identifies antibodies to alpha-gal. The woman contacted Diane Cymerman, an allergist she had seen years earlier for seasonal allergies. Cymerman had her patient list all the foods she consumed before her last episode in the hospital and had her blood tested for antibodies to everything on the list, down to the black pepper and parsley seasoning. And to alpha-gal.

The first results came back the same week; she had a moderate level allergy to beef but everything else was normal. The following month, the test results for alpha-gal antibodies came back. She was wildly allergic to galactose-α-1,3-galactose. Cymerman called the patient with the news. She had to avoid eating meat from mammals—and everything derived from them, including Jell-O and other foods and medications made from gelatin. Even foods cooked on a grill used for meat can be contaminated with enough alpha-gal to trigger a reaction.

The patient contacted the allergist back in Iowa and told him what she had. He was amazed. He had only recently heard a lecture on this phenomenon. He'd never seen it before her case.

It hasn't been easy for this Iowa transplant to give up eating beef and other meat that comes from mammals. Some days, she tells me, just thinking about a juicy hamburger or steak makes her stomach growl. But then she remembers her terror and that long drive to the Iowa hospital and sticks with chicken, fish, and vegetables.

Overflowing

They weren't looking for a diagnosis, the middle-aged woman explained. Her husband had a diagnosis. They just wanted help figuring out why, even with all the treatments he was getting, he wasn't getting better.

Until a year and a half earlier, her fifty-four-year-old husband had been perfectly healthy. Never missed a day of work, never so much as took an aspirin. Then he got what he thought was the flu. But even after the fever and congestion went away, the terrible body aches remained. He coughed constantly and felt so tired that just walking to the mailbox would leave him panting for air and shaking with fatigue.

Still, he went back to work. He enjoyed his job driving a locomotive for a manufacturing plant in rural Connecticut. Besides, his wife told me, he was a guy who needed to be busy. And he stayed busy until, one morning a couple of weeks after starting back at the job, he was driving to work and suddenly found himself rumbling over the road's shoulder, drifting toward the strip of grass and the woods beyond. All he remembered was that one minute he was on the road, and the next he wasn't. For the first time since he initially got sick, he was worried.

He went to an urgent care center. A chest X-ray showed fluid surrounding his lungs. And his EKG was abnormal. The nurse who saw the man was concerned. She ordered an antibiotic to treat a possible pneumonia and referred him to a cardiologist.

That was the first of many doctor's appointments but the last one he went to alone. His wife was worried that her husband, a quiet man who seemed to be wasting away in front of her eyes, wouldn't ask the questions that needed to be asked.

The cardiologist ordered a second chest X-ray, which showed even more fluid in the sac surrounding his lungs—so much that it was hard for the man to take a deep breath. The cardiologist sent him to have the sac drained of more than a liter of a clear yellow fluid. It made the patient feel better, but it didn't last; within days the shortness of breath returned.

Two weeks later, when the man saw a pulmonologist, that doctor referred him to have another liter of the same yellow fluid drained from his lungs. The man's abdomen began to swell with even more fluid. Where was it all coming from? No one could tell him. After the fluid reaccumulated just as quickly, his doctors sent the man to the hospital.

There, an echocardiogram—an ultrasound of the heart—showed that he now had fluid in the pericardium, the sac surrounding his heart. The heart could barely pump. He was rushed to the operating room, and a hole was cut into the pericardium to allow the fluid to drain and give his heart enough room to beat normally. A retinue of subspecialists searched for an explanation for this flood of fluids. His heart was strong, he was told. His lungs were fine. His liver was fine. There was no infection, and no cancer.

Finally, a rheumatologist found an answer. The patient tested positive for something called Sjogren's syndrome. In this autoimmune disorder, white blood cells attack the organs that make the fluids needed to lubricate the body. Those with Sjogren's often have dry eyes because they don't make enough tears, or a dry mouth because they don't make enough saliva. They have problems with dry skin, as well as with their joints and GI tract.

The patient and his doctors were thrilled to finally have a diagnosis. Still, most with Sjogren's don't need treatment for the disease itself, but can be treated for the symptoms of dryness and discomfort it causes. It was strange that the syndrome would, in this patient, produce this extraordinary fluid overload. The rheumatologist theorized that a second disease called undifferentiated connective tissue disorder (UCTD) might be involved. The patient was immediately started on two immune-suppressing medications.

He continued to need to have his lungs and abdomen drained of ten to twenty liters of fluids every couple of weeks. A third and then a fourth drug were added. But when, after months of treatment, he was no better, his wife insisted they get a second opinion. He was referred to a rheumatologist in New York. That doctor suggested still another immune-suppressing agent.

The patient was on four of these medications when I first met him that summer. Despite the medications, he continued to have liters of fluid drained from his belly and around his lungs.

After hearing about their terrible journey, I examined the

patient carefully, trying to find some clue as to what could be going on. His arms were thin and wiry, just bone and sinewy muscle; the overlying skin hung loosely, reflecting significant muscle loss. In contrast, his abdomen was huge—the belly of two Santas. The skin there was drum-tight. His neck, like his arms, was thin, and the veins on each side were hugely distended with blood.

Once he had dressed and I was able to gather my thoughts, I told the couple that only the heart could cause such a huge buildup of fluids. No, the man said emphatically: My cardiologist assured me that all the tests show that my heart is strong. I told them I'd pore through the thick folder of carefully organized records they had collected and come up with a plan.

I didn't believe it was his autoimmune disease causing all this. Even though he had Sjogren's and possibly UCTD, he was being treated. And when a remedy fails, you must consider the possibility that what it's treating is not the cause of the problem, and ask: What else could this be?

So I dug and thought and came up with a list of rarities that could cause these symptoms. I put the question to my friend and mentor at Yale, Andre Sofair, an internist on the faculty of the program where I had trained and now taught. His answer was familiar—surely this was the heart. I told him what the patient told me, that his heart had been tested and was in the clear. Andre was surprised but turned his mind to thinking of other causes. He added a couple more items to my list.

I sent the patient for more tests, and when they came up with nothing, I thought back to Andre's first instinct. Was it his heart? In such a setting, I think of the Russian proverb

Ronald Reagan used during his negotiations with Mikhail Gorbachev on the treaty regulating intermediate-range missiles: One must trust, but also verify.

When the heart muscle has been damaged by, say, a heart attack, it doesn't pump as well, and fluids can back up. We call that congestive heart failure, and it was one possibility. But the bulging neck veins I saw suggested another, rarer possibility: constrictive pericarditis. In this disorder, the pericardium is injured—usually by a viral infection—and as it heals, it shrinks. Stuck in this shrunken jacket, the heart can pump only a fraction of the blood needed by the body. Could the virus that caused the flu-like symptoms at the start of this illness have attacked his pericardium?

I sent the patient for another echocardiogram. It showed a heart pumping hard but constrained inside a shrunken, thickened pericardium, unable to process the normal measure of blood. I spoke with his rheumatologist, who stopped all the immune-suppressing agents, and I sent him to John Elefteriades, a well-respected heart surgeon at Yale. Elefteriades cut away the damaged sac. Once he made the initial incision down the length of the scarred pericardium, the blood flow through the heart more than doubled.

The man's recovery from this surgery has been remarkably fast. He had a foot-long incision down the middle of his chest, but within two weeks of the operation he was home and walking around. Three weeks later he was back at work—just in time for him and his wife to prepare for a holiday season they had worried they would never see together.

Muscle-Bound

The young teacher paced between the rows of teenagers. It was his second day on the job, and he was nervous. His heart was pounding. His tie felt unbelievably tight. Suddenly, it was hard to breathe. Really hard. He could feel sweat beading coolly on his face. He glanced at the clock. Could he make it to the end of the period? Finally, the bell rang—class was over.

The hallway to the nurse's office seemed to stretch out into the distance. He could feel himself go through the motion of breathing, but the air didn't seem to make it to his lungs. "I can't breathe," he croaked, leaning against the door of the medical office. The school nurse led him to a bed. He could hear her asking questions, trying to get more information, but it was hard to speak. She removed his tie, then placed a mask over his mouth and nose. The cool rush of oxygen brought some relief from the sense of drowning on dry land. The next thing he remembered was being loaded into an ambulance.

At the hospital, doctors diagnosed a massive pulmonary embolism, which occurs when part of a blood clot breaks off and is carried through the circulatory system into the vessels of the lungs. In this patient, it was a very large clot, which prevented most of the circulating blood from reaching his lungs, where the oxygen he breathed could be exchanged. He was started on blood thinners and admitted to the ICU. As soon as he was stable, the doctors turned their attention to the clot itself. Where did it come from? Why did it form? They needed to find out because another assault like that could kill him.

Our lives depend on our ability to form blood clots. But like so much in the body, context is everything. In the right place, at the right time, a blood clot can save your life by preventing uncontrolled bleeding. In another setting, that same clot can kill. Clots normally form at the site of injury to a blood vessel. They can also form when blood stops moving; that's why anything that causes prolonged immobility, like traveling or being stuck in bed, increases the risk of a clot. Certain drugs—estrogen and other steroid hormones—can also increase the risk. Some people have a genetic abnormality that makes their blood coagulate too readily. Finding the cause of a clot is crucial to estimating the risk of another.

So the patient's doctors looked. They found nothing in his legs—the most common source of pathological blood clots. CT scans of his chest, abdomen, and pelvis likewise showed nothing. He hadn't traveled recently, hadn't been sick. He took no medicines. His doctors sent off studies of his blood to look for any evidence that his blood was too eager to clot. Normal. They could find no reason for this otherwise healthy young man to have developed a clot.

It's difficult to be a patient with an illness that can't be explained. It's even harder when a diagnostic uncertainty leads to an unacceptable therapeutic certainty: In this case, the patient was told that he would have to take warfarin, an anticlotting drug, for the rest of his life. He was twenty-three years old and liked playing sports. He played baseball and basketball in high school and rugby in college. But when you can't clot, these games become dangerous. The drug would protect him from another pulmonary embolus, but in return he would have to avoid anything that could cause bleeding.

The patient hoped for an alternative and found Dr. Thomas Duffy, a hematologist at Yale University with a reputation as a great diagnostician—the kind of doctor that other doctors turn to when they are stumped. Perhaps this doctor could figure out what caused his devastating pulmonary embolus and help him get off warfarin.

Duffy is a tall, fit man in his sixties with a preference for bow ties and a precise, thoughtful manner of speaking. He listened to the patient's story, then asked for a few more details. What kind of physical activity had he been doing in the weeks before the clot? He lifted weights every other day and either ran or swam the days between. Had he taken any performance-enhancing drugs? Yes, but not for years.

Duffy considered the possibilities. The usual suspects had already been ruled out; whatever caused this clot was going to be uncommon. Could a clot have formed inside one of his organs—his heart, his liver, his spleen—and traveled to his lungs from there? The scans the patient had wouldn't have shown that. A myxoma, a rare type of tumor that grows in heart muscle, can cause a clot within the heart itself. Could he have such a tumor? An uncommon blood disease called paroxysmal nocturnal hemoglobinuria can cause blood clots in the liver, in the spleen, or beneath the skin. Did he have this rarity? The physical exam might give some clues.

When the patient undressed for the exam, Duffy was immediately struck by the highly developed muscles of his upper body. "He looked like one of those young men in a men's fitness magazine," he told me later. Otherwise his exam showed nothing abnormal.

Then Duffy remembered a physical-exam maneuver he learned years ago when he was a medical student. He straightened the patient's arm and held it parallel to the floor. Carefully placing a finger over the pulse at the young man's wrist, he moved the arm behind the patient. Then he asked the patient to tilt his head up and face the opposite direction. The pulse disappeared. When the patient looked forward again, the pulse returned. He repeated the maneuver. Again, the pulse disappeared when the patient turned his head. Immediately Duffy suspected what had caused the clot.

The vessels that carry the blood from the heart to the shoulders and arms and then back to the heart must travel through a narrow space under the clavicle and above the top of the rib cage. The presence of hypertrophied muscles of the shoulder or neck, or in some cases an extra rib, can make this small opening even tighter. This problem, known as thoracic outlet syndrome, is most commonly seen in young athletes who use their upper extremities extensively—baseball pitchers or weight lifters—or in workers who use their arms above the level of their shoulders, like painters, wallpaper hangers, or teachers who write on a blackboard. When these patients put their arm in certain positions, the extra muscle or bone constricts the space between the two structures and cuts off the flow through the vessels like a kink in a garden hose.

Blood can't get into the arm, so the pulse disappears. And blood can't get out of the arm, so it pools and can clot. When the arm is moved and the vessel reopens, the blood flows once more, but if a clot has formed, it can break loose and travel to the lungs.

Duffy ordered additional tests to make sure there was no other explanation for the clot. He then referred the patient to a surgeon who had experience with an unusual and difficult surgery: the removal of the first rib to widen the narrow opening. Nine months later, during the summer, the patient had the surgery. Three months after that he was able to stop taking warfarin. Four years later, the patient was teaching, playing sports, and lifting weights without difficulty.

"Seeing that extraordinary musculature reminded me of this unusual anatomical abnormality, and the test, of course, I learned many years ago," Duffy recalled when I spoke to him about the case. I had never heard of the old-fashioned arm maneuver. This and other physical-exam techniques are part of a disappearing tradition in medicine—replaced more or (in this case) less successfully with a variety of high-tech imaging techniques. Yet had a doctor not done this simple test, the patient's abnormality may not have been picked up, and he would have been stuck taking medicine he didn't need and missing out on the sports he loved.

Uphill Battle

The patient was halfway up the stairs by the time he noticed how short of breath he was. When he reached the top of the staircase, he had to stop, sit down, and catch his breath. That had never happened before. "It scared me," he told the middle-aged doctor to whom he was describing his symptoms—the third doctor he had seen since that day on the stairs. "I never get sick. I didn't feel sick. I just couldn't breathe."

He was a healthy man in his fifties who rarely saw his primary care doctor, but that day he wanted to see him—right away. When his doctor asked him about his symptoms, he had few to report. No cold symptoms, no fever or chills, no joint pains. He hadn't lost weight; he wasn't tired. But when he did anything active, anything at all, he felt as out of breath as if he had run a fifty-yard dash. He had no other medical problems except high cholesterol, and he took Lipitor to keep it down. He had never smoked. He drank socially, was active, though not athletic, and had his own law practice.

He looked younger than his fifty-nine years, with serious hazel eyes and an easy smile, but the doctor noted that he started breathing hard while walking to the exam room. By the time he boosted himself onto the table he was sweating. Other than his rapid breathing, his exam was pretty normal. With one exception: In the bottom half of both lungs, the doctor heard a quiet but abnormal sound, like Velcro being pulled apart with each breath.

The patient's history and lung exam suggested that he probably didn't have pneumonia. There is usually a history of fever and cough with pneumonia, and on examination the most common finding is a long snore-like sound deep in the chest or, more ominously, that the affected region of the lungs is quiet, with few of the normal sounds of moving air. Because of the sudden development of his symptoms and the fact that they worsened when he exerted himself, the doctor worried that the problem may have been his heart, not his lungs.

The patient was the right age for heart disease and had a history of high cholesterol. The doctor sent the man to a nearby cardiologist for evaluation. He also ordered a chest X-ray. The cardiologist gave his heart a clean bill of health. A stress test showed no evidence of any reduced blood flow to the heart, and an echocardiogram showed that his heart was beating normally.

The chest X-ray, however, was not normal. At the base of each lung there was an infiltrate—an area of lightness where there should be darkness—suggesting that something other than air and delicate lung tissue was present. Pneumonia is the most common cause of such a finding, but again, the patient had no other symptoms of that kind of infection.

After the cardiologist cleared the patient's heart, his internist tried a week of antibiotics. When that had no effect, he sent him to a local lung specialist, who tried a week of steroids. When that also brought no improvement, the patient sought out Dr. Charlie Strange, a pulmonologist at the Medical University of South Carolina. Strange listened to the man's story, examined him, and reviewed the chest X-ray and the CT scan that had already been done.

Then he listed what he thought were the most likely causes of the patient's symptoms. First, it could be an infection. Although the patient had been treated with a good antibiotic, there were many unusual organisms that wouldn't be affected by most antibiotics. Another possibility was an inflammation of his lung tissues, what's known as interstitial lung disease. This category of disease includes many uncommon disorders that all cause shortness of breath and lung damage. While there are more than a hundred of these diseases, Strange explained, what really matters is that many of them respond to steroids.

However, a fraction of them progress despite all therapy; patients with these untreatable forms of the disease lose more and more lung tissue over time and die—usually within a few years of their diagnosis. Lung cancer was also a remote possibility. The patient had never smoked, and the infiltrate was in both lungs. Those two factors argued against most lung cancers. But there are rare types that can spread rapidly or start in more than one location.

Strange sent off blood to look for evidence of an unusual infection. Other blood tests were performed to check for diseases known to cause interstitial lung disease. The patient was also scheduled for a bronchoscopy—a specialized test in which a fiber-optic scope is sent through the nose or mouth deep into the airways. Cells and tissues taken from the lungs could reveal an infection or a cancer.

There was no evidence of either when the tests came back the next day. That meant that the patient had an interstitial lung disease. But which one? It would take an open biopsy—surgery—to get a full diagnosis. Perhaps the more important question was, would the disease respond to steroids or would it progress and kill the patient within the next few years? The patient had tried steroids earlier but stopped after only a week because he had felt no better. It's possible that those few days weren't enough to get a response.

The bronchoscopy did provide one clue suggesting that the patient's disease might be treatable: in his lungs there was a high level of a single type of white blood cell—eosinophils. These cells are most frequently associated with allergic reactions; when seen in interstitial lung disease, they usually mean that the patient will respond to steroids, a potent anti-inflammatory.

"What medicines are you taking?" Strange asked. Only Lipitor, for cholesterol, the patient reported. "Stop the Lipitor and start the steroids," Strange instructed. "Let's see if the Lipitor is part of the problem."

At home, the patient started his own test: he monitored how many breaths it took to get back to "normal" after climbing two flights of stairs. Day 1, the first day off Lipitor and on the prednisone, it took him one hundred breaths. Three days later, it took him fifty-five breaths. Three days after that, twenty-five breaths. A week later, he was down to eight breaths.

He triumphantly announced his success when he saw

Strange. "I'm not a hundred percent, but I'm better. Much better." The pulmonologist agreed; he did look better. He measured the patient's oxygen saturation. Sitting, it was a little low, 94 percent. Walking, it dropped to 85 percent. Normal is 100 percent sitting, walking, even running. His breathing was improving, but his lungs were still damaged.

It would be years before the patient would know how much lung he could recover. And he never found out which interstitial lung disease made him ill. Strange narrowed it down to two possibilities, but the treatment for both is steroids, and the patient was already on them, so there was no reason to go through the surgery and hospital stay to narrow it down further. Years later, he still is not willing to go back on medication to lower his cholesterol. He and his doctor suspected that an allergic reaction to Lipitor triggered his illness, though they could never be sure. He remained optimistic about the future, and told me: "If it's a long, slow total recovery, well, I can live with that."

Brokenhearted

"I'm not going to lose my mom." The young man's voice cracked with feeling. Beside him a half dozen men and women in scrubs swarmed around the gurney rolling the woman into a cubicle in the ICU of the J. W. Ruby Memorial Hospital in Morgantown, West Virginia. The patient's face had a deathly pallor, her light brown hair was dark with sweat, her mouth was open, and her chest heaved as she struggled to breathe. "We'll do our best," the doctor assured him as he observed the woman and the monitors that showed just how sick she was. The young man, who was in his midtwenties, grabbed the doctor's arm as he turned to follow the patient. "No—you have to save her," he answered fiercely. "You have to."

His mother had been fine that morning, the young man told the doctor. She went to work just as she did every day. But then the phone rang, and she learned that her husband of more than two decades had been killed in a crash. She rushed to the site, found his body, and collapsed next to him, sobbing and shouting his name as if she were trying to wake him. She lay next to her husband, cradling him in her arms until his body was taken away. Two hours later, she collapsed again, and this time she couldn't get up.

The son paused and roughly rubbed the tears from his face with his sleeve. When his sister got home, his mother told her that her chest hurt and that she felt as if she couldn't breathe. The ambulance rushed her to the nearest hospital. "The doctors there told us she'd had a heart attack," the young man continued, "and that she was fixing to have another one." He and his two sisters were terrified. They had already lost their dad—they just couldn't lose them both. They decided to have her transferred here, to the regional hospital with a specialized cardiac-care unit.

The doctor in the ICU glanced through her chart and turned his attention to the patient. She was forty-five years old and a smoker. She had recently been told she had narrowing of the arteries that carried blood to her legs and feet—called peripheral vascular disease—but otherwise she was healthy. She took no medicines and worked full time now that her children were grown up.

On examination, she appeared younger than her forty-five years. But her tanned, unlined face was shiny with sweat, and her pale blue eyes were open and unfocused. Her heart was beating rapidly, and a blood-pressure cuff that inflated automatically beeped its warning that her pressure was dangerously low. An oxygen meter on her finger showed that although she was breathing rapidly, she wasn't getting enough air. Her skin was clammy to the touch and pierced by thick intravenous catheters delivering saline and medicines to raise her blood pressure.

It was clear that the patient's heart was failing. She was young to be having a heart attack, but she was a smoker and had a history of clogged arteries in her legs, which put her at risk of having the same problem in her heart. A heart attack occurs when one of the arteries supplying blood to the heart gets blocked. Without blood, that part of the heart dies rapidly. Her EKG was abnormal, and blood tests revealed damage to the heart cells—all consistent with a heart attack.

Dr. Conard Failinger, the cardiologist on call, was worried by the grainy images of the sonogram that showed the patient's heart in motion. Her heart was pumping with only a fraction of the expected strength. In fact, most of the heart muscle wasn't pumping at all; the patient was dying. The only way to treat her would be to quickly find and clear the blockage so that blood could flow once more. There are chemical clot busters that can do this, but a more effective way is to thread a tiny catheter into the affected artery, locate the blockage, and then use the tiny tube to blast the vessel open. Done quickly enough, this process, known as cardiac angioplasty, can save the heart muscle and save the life. The patient was quickly transported from the ICU to the "cath" lab.

Once there, Failinger watched another doctor rapidly thread the tiny catheter through a large artery in the patient's leg into her heart. He carefully placed the catheter into one of the major vessels of the heart and pressed the plunger of the attached syringe, which shot a tiny amount of contrast dye into the artery to determine the site of the blockage. On a monitor, the cardiologists stared in wonder as the arteries brightened, lighted by the dye flowing through them. There was no obstruction. The doctor manipulated the catheter again, moving it to another vessel. Again, the dye flowed through the artery, completely unimpeded. Several more tries produced the same result. There were no blocked arteries. The patient was not having a heart attack.

What else could cause such profound heart muscle weakness? Alcohol can do this, but the patient had no history of heavy alcohol use. A number of drugs—most commonly those used to treat certain cancers—can cause this type of damage, but this patient had never been exposed to any of those medications. Infection could do this, but the patient reported no symptoms other than those caused by the failing heart itself.

Failinger immediately realized that it was none of these. He recognized what it was, though he had never seen it. He had read about it a short time before in *The New England Journal of Medicine*. This was stress cardiomyopathy, also called "broken heart syndrome." First described by the Japanese in 1990, this disease occurs when an emotional trauma causes the brain to release high doses of stress hormones. This hormonal blast paralyzes the muscle cells of the heart, preventing them from working to pump the blood. Typically only one section of the heart is spared this devastating paralysis—the part closest to the aorta so that with each heartbeat only the upper portion contracts and the heart looks like a narrow-necked vase. The Japanese called it *takotsubo*, after a type of trap that is used to capture octopus and has the same vase-like shape. For reasons that no one understands, this mostly affects postmenopausal women.

There is no cure. There is no clot to bust, no bugs to kill. Like its metaphorical counterpart, the only treatment is support and the passage of time. The initial burst of hormones subsides and the patient must be kept alive until the heart recovers. For those who survive long enough to reach the hospital, the prognosis is good. Once she made it to the hospital, this patient needed additional oxygen and medicine to keep her blood pressure up. When she arrived, her heart was able to pump out only 5 to 10 percent of the blood it contained (normal is 50 to 60 percent). After several days, it was pumping well enough for the doctors to stop the medications that increased her blood pressure. By the end of the week, her heart's capacity had doubled. Just days later, it was nearly normal.

"If anyone had told me that you could die of a broken heart," the patient told me recently, "I'd never have believed it. But I almost did." Reflecting on those couples you hear about, when one dies and the other one follows a few days later, she said, "I bet their hearts were broken, just like mine was."

It's an interesting idea: Perhaps the metaphors we use to speak of devastating loss grew out of physiological truth. But if love lost almost took this patient's life, she says she believes that it was also love that brought her back. "I remember when I was in the hospital, I was in the most peaceful rest," she told me. "I didn't see any light or anything, but it was just as beautiful and peaceful a rest as I could ever imagine. I just wanted to stay there forever. But then, way off in the distance, I heard my kids calling to me, and I knew I couldn't stay. They're the ones who really saved my life."

Deflated

"Are you okay?" the man asked his wife. It was two A.M., and he'd awakened to find the bed next to him empty. He found his forty-five-year-old spouse in the living room of their weekend house up the Hudson River from New York City. She had an oxygen meter on her finger and a worried look on her face. "I can't breathe," she told him. She'd had chest pain and felt short of breath in the past, but her oxygen had never been this low—down to 89 or 90 percent. And the right side of her chest felt as if it were on fire.

She wanted to tough it out until morning so that they and their two young daughters could drive back to Manhattan where her doctors were—the ones who had been treating her since her right lung collapsed and all the trouble began two years earlier. That first time had been nothing like this. Back then, there was a strange click, then a feeling that something had moved. It wasn't painful, just odd. A few days after that, she developed a cough. Her doctor thought she had a virus.

When she got worse, he prescribed an inhaler. A few more days passed, and she felt out of breath just walking to the bathroom—odd for a woman who usually exercised daily. Her doctor ordered a chest X-ray and saw to his surprise that her right lung had collapsed.

She had, he explained, a pneumothorax—literally, air in the chest. It happens when there's a tiny rupture in the lung. The air rushes out into the surrounding space, and the empty lung collapses. She was admitted to Lenox Hill Hospital the next day. A tiny catheter was inserted between her ribs into the space around the lung. The air was sucked out, allowing her lung to reexpand.

But why had she gotten this leak in the first place? She didn't smoke, which is the most common risk factor for developing a pneumothorax. None of the tests indicated any type of lung disease, another significant risk factor. And though there are a number of inherited diseases that can predispose a person to developing a pneumothorax, no one in her family had any of them. After four days and no answers, her doctor concluded that she'd had a spontaneous pneumothorax. These are rare but are more likely to be seen in tall, thin, athletic individuals—like her—and don't usually happen more than once.

But a year and a half later, while at work, she felt that same immediately recognizable click and shift in her chest. An X-ray revealed another pneumothorax. At Lenox Hill Hospital, her lung was again reexpanded, and Dr. Byron Patton, a thoracic surgeon, recommended a procedure known as pleurodesis, in which the lung is mechanically attached to the surrounding sac, called the pleura, so that even if the lung developed another leak, it wouldn't collapse.

Over the next half year, the patient continued to have

episodes of chest pain on her right side. Each time she had a twinge or stab, fearing another pneumothorax, she would call her doctor, and he would send her to the ER for an X-ray. It happened almost twenty times. The X-rays, though not exactly normal, did not show evidence of new leaks—until one day in May, when a chest X-ray showed yet another, albeit small, pneumothorax.

Why was this happening? The patient asked Patton whether the lung problems could be related to the hormone therapy she'd had for in vitro fertilization (IVF). It was really the only new thing in her life before all this started. It had taken nine fertilizations and nearly three years, but she finally became pregnant four years earlier and now had twin girls. Patton didn't know of any link between pneumothorax and IVF, but there was a link, albeit rare, between pneumothorax and endometriosis—a disorder in which tiny dots of the inner lining of the uterine tissue, known as the endometrium, move into other parts of the body. Women who have endometriosis can have episodes of pneumothorax if those cells make their way past the diaphragm into the chest.

This endometrial tissue, like that in the uterus, changes with the monthly cycle of estrogen and progesterone, causing pain and sometimes bleeding. In the chest cavity, it could cause what was called a catamenial (derived from the Greek for "monthly") pneumothorax. But did her episodes of pneumothorax come with her period? She wasn't sure.

Patton had suggested starting birth control pills to suppress the hormonal changes. She began taking them three weeks before the horrible night at her weekend home, where she woke up short of breath. First thing the next morning, her husband carried their sleepy children to the car. She carried the oxygen tank she got earlier that spring—just in case—and they set off for the drive to Lenox Hill Hospital. They stopped to pick up her mother on the way.

At the hospital, an X-ray showed that her right lung, which just weeks before had a small pneumothorax, had now collapsed completely. Patton was shocked. The pleurodesis hadn't worked. The patient had been reading about catamenial pneumothorax, and even though hormone suppression hadn't been effective, she still thought this was what she had. Her mother found a gynecologic surgeon at Lenox Hill, Dr. Tamer Seckin, who specialized in the diagnosis and treatment of endometriosis. Mother and patient urged the two subspecialists to work together in the operating room, and they agreed.

The operation, which involved both doctors, took five hours. Seckin went first. He looked in the abdomen and pelvis and found many endometrial implants. He saw small patches of wayward cells on the bladder and the intestines and scattered across the abdominal and pelvic walls. But he did not see any on the bottom side of the diaphragm, the muscle that separates the abdomen from the chest. Although there was evidence that she had extensive endometriosis, it wasn't clear if it was the cause of her lung collapse.

Then it was Patton's turn. He would examine the chest and lungs. The patient was placed on her left side, and the camera and surgical instruments were inserted between her ribs into the right side of her chest. Patton carefully examined the smooth, curved upper surface of the diaphragm. Directly below the lung there was a patch of purple tissue, a little smaller than a dime. Were those the endometrial cells? Patton cut out the abnormal-looking tissue, then sewed up the hole. He scanned the lung, from top to bottom, and found another small patch of abnormal-appearing tissue, which he cut out as well. The samples were sent to the pathology lab. He then fastened her lung to the pleural surface of her chest again. Before the operation was finished, the results were back from the lab. The tissue on the diaphragm was endometriosis. She had catamenial pneumothorax disease.

Recovery was slow after the operation, but the patient felt less anxious because she knew the cause of her lung problems. In a second operation, Seckin was able to remove all the visible endometrial implants. Still, there was no way to guarantee that she wouldn't develop more unless the source of the implants, her uterus, was removed, along with her ovaries. It was a tough decision. But since she wasn't planning to have more children, she felt it made sense.

Six months after that final operation, the patient said her life was slowly coming back. She joked that after all those chest and abdominal operations, she looked as if she'd been in some kind of knife fight. And though her chest still felt tight, it was good to be able to breathe.

PART V

All in Your Head

Honeymoon in Hell

"Something's wrong," the twenty-seven-year-old woman said to her new husband. "I think you need to take me to the hospital." It was the day after their wedding. The woman's husband and her best friend were car fanatics, and so the newlyweds had wanted to commemorate their union with pictures at a drift track in rural Toutle, Washington. The best friend would "drift cookies," circling the couple in a tight, controlled skid. As another friend took pictures, the two embraced, wreathed by smoke and dust and barely contained chaos as the red Mustang fishtailed around them. In the photos, the couple look happy.

But as they loaded up the car to go home, the young woman started to feel strange. She'd been a little jittery all day. She noticed she couldn't stop talking. She figured it was just the excitement of the wedding and its aftermath. But suddenly her excitement felt out of control. Her heart, which had been racing since she got up that morning, went into overdrive. It pounded so hard that it hurt her throat and chest. She couldn't think. Her hands took on a life of their own—they opened and closed incessantly.

Her new husband was confused and worried. They drove to a hospital a couple of towns over. It was a panic attack, the doctors there told the couple. Since the birth of their daughter the year before, the young woman had struggled with postpartum depression and anxiety. She'd just married and had these crazy pictures taken; it was no wonder she was panicking. The young woman accepted the diagnosis, but she couldn't help feeling that this was very different from the anxiety she sometimes experienced.

She was given a medication to take if she had more symptoms and sent home. The pills didn't seem to help. The next day she felt her heart pounding in her throat and the same spacy-headed jitters from the day before. She tried the medicine again, but after that, her memory is just fragments.

She doesn't recall her many trips to the emergency room over the next few days. It was clear that something was very wrong, but the doctors didn't have answers beyond anxiety and depression. When she started speaking nonsense in a strange babble, doctors added a new term: psychosis.

After a week of repeated visits to emergency rooms, a social worker suggested she go to Telecare, a psychiatric hospital in Vancouver, Washington. After more than two weeks there, doctors were concerned that this wasn't a psychiatric illness after all, and she was transferred to the nearby PeaceHealth Southwest Medical Center. After three days of evaluation there, no medical cause for her symptoms was found, and the patient was admitted to the hospital's psychiatric unit.

The psychiatrist there thought her diagnosis was something known as excited catatonia. Catatonia is usually defined by a slowing of movement, thought, and speech. Excited catatonia is much less common, and its defining characteristic is agitation. But the excitation isn't seen just in actions or words. It can progress to a point where blood pressure and body temperature rise to life-threatening levels. Both forms of catatonia typically respond immediately to low doses of a type of sedative known as benzodiazepines.

This young woman was certainly agitated. But Dr. Michael Rothenfluch, the psychiatrist in charge of her care, thought that she didn't look like the cases of excited catatonia he had seen in his seven years. The woman had three symptoms that were atypical: She was confused; her speech was garbled; and she seemed to have seizure-like episodes of shaking and inattention. In addition, she was not responding to the medication. He was worried about her and asked his senior colleague Dr. Michael Bernstein to see her.

Bernstein visited the patient later that day. She was in the locked part of the ward, where the patients at greatest risk to themselves were placed. The young woman was lying in bed with her eyes closed—naked and disheveled. The sitter assigned to stay with her kept trying to cover her with a sheet, but she repeatedly tossed it off. She moved constantly and restlessly on the thin mattress. The bed was the only piece of furniture in the room. Light streaming through the scratched window illuminated the otherwise bare space, which was designed to keep the occupant from hurting herself or anyone else.

Bernstein squatted next to the patient and gently spoke her name. She opened her eyes but didn't look at him. He asked how she was doing. Okay, she mumbled. He asked, "Can you tell me more?" No answer. "Do you know why you are here?" "Call my parents," she said. "Do you know where you are?" he asked. "Call my parents." Suddenly she began to gag as if she was trying to speak but her own body was trying to stop her.

Excited catatonia is usually seen in those with a long history of mental illness. Rothenfluch called the patient's mother. Had she ever had any symptoms of mental illness in the past? Never. She had been a normal girl with the usual ups and downs. She was depressed after her baby was born but nothing like this.

After some time on the benzodiazepines, the patient was still no better. The two psychiatrists discussed other possible causes of her psychiatric symptoms. Bernstein recalled patients he'd seen in the past with psychosis that resulted from a tumor. These so-called paraneoplastic syndromes can be caused by substances secreted by the tumor or an immune reaction of the body to the tumor itself.

Before her transfer to the psychiatry unit, a neurologist at PeaceHealth had considered the possibility of a paraneoplastic syndrome triggered by antibodies to an ovarian growth, known as a teratoma. These tumors contain a mixture of cell types—bone, skin, muscle, and tissue from various organs. But on rare occasions, the tumors will grow some type of brain cells. These cells prompt the body to develop antibodies that will attack and destroy the same type of cell in the brain. The neurologist decided it was unlikely that she had a paraneoplastic syndrome because she didn't have the expected changes on her electroencephalogram. In addition, brain lesions from paraneoplastic syndromes can sometimes be seen on an MRI. The patient had had these scans and they were normal.

And yet as the psychiatrists considered the various paraneoplastic disorders, psychosis triggered by a teratoma was a possibility. Many of the other disorders are visible on brain imaging and the patient's MRI had been normal. Moreover, it was a disease most commonly seen in young women, like this patient. They ordered the blood test to look for the specific antibody seen in this disease.

Six days later, the results came back: She *did* have a paraneoplastic syndrome, which turned out to be triggered by a teratoma. A CT scan located a walnut-size tumor on her ovary.

She was transferred back to the hospital's medical team. The tumor was removed, and the level of the antibody started to drop. Recovery, however, wasn't nearly as fast. The repair of the cells damaged by the attacking antibodies took time. She went home to her mother's house. Her husband, a logger, worked outside the state and visited as often as he could, but they couldn't afford to have him quit his job to help her. It has been six months, but she still has some issues with her memory.

Bernstein was amazed by the diagnosis. Teratomas that trigger this type of brain destruction are rare and only recently described; the first two cases of psychosis caused by a teratoma were reported just twenty years ago. This case made him wonder about two other young women he cared for decades earlier who died from brain disorders that started out looking like psychotic disorders. Could those symptoms have been caused by this kind of then-unknown antibody? He feels certain of one thing: He and his colleague will always be aware of this diagnosis, should they ever see it again.

A Different Man

So, do you like working here?" the middle-aged man bellowed to the young physician at the other end of the hospital coffee shop. The woman, the object of this not-very-subtle pickup line, ignored him. The man's sister cringed. When had her younger brother turned into such a jerk? He had always been so quiet and shy. She was living across the country in Washington State, so she didn't see him often, but he had certainly changed.

In his twenties, he had a problem with alcohol. But back then, his drinking made him quieter. And even during the worst of his drinking days, he had always been tidy and fastidious in everything he did. That morning, she drove from the airport to pick him up on the way to visit their father, who was in the hospital after heart surgery. She had taken a red-eye from Seattle to Philadelphia, but her brother looked worse than she felt: tired, disheveled, dirty. He said he had just showered, but she could tell it wasn't true.

As their father recovered, she had plenty of time to discuss her brother with the rest of her siblings—three sisters and another brother. All had noticed that something was different—how could they not? He would say the most inappropriate things, and always loudly. He slept often and everywhere. And he couldn't remember anything. One sister said she was pretty sure he was drinking again. That produced a couple of nods of agreement.

The younger brother insisted that he was fine, even as he got worse. He didn't know why he was no longer able to get work as a sheet-metal mechanic. He didn't tell them that he had often forgotten important tasks on the job or that he had gotten lost more than once on a construction site where he was working.

Nine months after their father's surgery, the sister who lived the closest made plans with her brother to have brunch. Early that morning, he called to confirm the date and the time. And then he called again, and again, and again. More than a dozen times. When she arrived to pick him up, he wasn't dressed and seemed to have completely forgotten their plans—and all his calls. She hustled him into the car and drove him to a local emergency room. She explained that her brother had become very forgetful and that his personality had changed significantly. The doctors drew some blood and ran a CT scan. All was normal. He should see a neurologist, she was advised.

The sister arranged for him to see Dr. Adam Weinstein, a young neurologist at the Center for Neuroscience in a suburb of Philadelphia. At that first appointment, the sister did most of the talking, Weinstein noted. He soon understood why.

The brother sat quietly in the exam room. He couldn't explain what had been going on for the past year. He said he was laid off from his job because he "couldn't focus." It wasn't because he was drinking—he hadn't touched a drop in five years. The man's face revealed no emotion as he spoke. It was hard for him to answer with more than a simple yes or no, as if his vocabulary had evaporated from his mind. And though he knew the name of the president, he couldn't recall what year it was, or even the day of the week.

As a neurologist, Weinstein saw many patients with dementia. The trick was to look for causes of cognitive impairment that could be reversed. They were uncommon but worth looking for, especially in someone this young. He would look for evidence of advanced syphilis, vitamin deficiencies, thyroid problems, and, because the patient had worked with steel, heavy-metal poisoning. Seizures could affect the brain, so he ordered an electroencephalogram, as well as an MRI, in case something was missed by the CT scan.

Blood-test results were unrevealing. The EEG showed that the man's brain was working slower than normal, but nothing suggested seizures. The MRI was scheduled to take place weeks later. If it didn't provide some answers, Weinstein knew he would have to talk with the family about how to deal with dementia in someone you love.

It was late in the afternoon when Weinstein got a call from the neuroradiologist reviewing his patient's MRI. It indicated a condition that Weinstein had read about but never seen: severe spontaneous intracranial hypotension, or "sagging brain syndrome."

Normally the brain floats in a bath of cerebral spinal fluid, which cushions and protects the delicate structure. In this disorder, that liquid disappears. Weinstein pulled up the images from the man's MRI. They showed the man's brain, which would normally have the consistency of a light and fluffy custard, lying in a pile on the floor of the skull. The whole brain seemed to sag downward through the hole from which the spinal cord emerges from the bony sphere. Weinstein could see why the man's memory and emotions were affected. And why he spoke so loudly. The temporal lobes—where sound is heard, memories are formed, and emotions originate—had been pulled back and down, stretched like saltwater taffy.

Weinstein immediately started reading up on the condition. It is usually caused by a leak in the tough sac surrounding the brain and spinal cord called the dura. Weinstein knew that the first step would be to find the leak. Once the opening was found, injecting the patient's own blood into that area would patch the hole temporarily, giving it a chance to heal.

He sent the patient to a nearby community hospital and ordered imaging to look for the leak—a procedure in which contrast dye is injected into the dural sac. If there is a leak, it becomes visible as the contrast oozes out of the dura along with the cerebral spinal fluid. Although the radiologists suspected that there was a leak, they couldn't find it.

The patient was transferred to a larger medical center. Doctors there had no more success in finding the leak, but they tried to treat him anyway. Three times his blood was injected into his spine. Each time he got better for a week or two, but then the leak returned and, overnight, all his improvements disappeared.

Finally, a doctor encouraged the family to contact a neurosurgeon on the West Coast who had recently developed an experimental technique for repairing these leaks. The surgeon, Wouter Schievink at Cedars-Sinai Medical Center in Los Angeles, had treated only a small number of patients, but he was one of few surgeons working on this strange and unusual disorder. Collaborating with neuroradiologists at Cedars-Sinai, Schievink had a better way to look for leaks and performed surgery to repair them.

The West Coast sister emailed Schievink her brother's radiology reports. Bring him out, the surgeon told the desperate sister. If he could find a leak, he would operate.

A few weeks later, the patient went to Cedars-Sinai. Schievink's team found the leak. Finding it had been hard because the contrast dye wasn't leaking into the surrounding space but instead draining into a spinal-cord vein. There it was quickly diluted and carried away. The very next day, the patient was taken to surgery. It took three hours to locate and close the pea-size hole.

Recovery was slow and brutal. The man's brain had adapted to the low pressure created by the leak, but once it was repaired and the pressure normalized, his brain and all the blood vessels feeding it had to readjust. For those first few days, his head pounded relentlessly. He threw up everything he tried to eat. But slowly, so very slowly, he started to improve. His family watched, amazed, as the brother they knew so well reemerged from the chaos of his dementia. After a month and a half, he was ready to go home. Four months later, he got the okay from his doctors to go back to his job.

He has been working ever since. Whe I spoke to him, he was looking forward to seeing his family for Thanksgiving. As he told me, he has much to be thankful for.

Unexpectedly Drunk

The radio crackled to life with a burst of static. It was a Saturday night, and the emergency department was packed. An EMT's voice silenced the harsh electronic noise: "We've got a thirty-five-year-old Caucasian man with an altered state of consciousness. His friends think he might have been slipped some drugs." Minutes later, a young blond man was rolled into the ER, screaming and shouting obscenities as he struggled against the straps that held him in place on the stretcher. "I want to go," he yelled. "I want to go."

The physician, a man in his fifties with gentle brown eyes and a neatly trimmed beard, approached the three men standing at the patient's bedside. "I'm Dr. Shavelson. Can you tell me what happened?" All three began speaking at once, then stopped. A young man began again: "He was fine this morning. I had lunch with him. Then he went to the sauna. He called me a few hours later and said he felt like he'd been drugged." He told the friend that he was hot, nauseated, and lightheaded; he was having a little trouble walking. By the time he got home, he felt worse, not better. So he called friends who lived nearby. He was having trouble seeing, he said, as if he were in a very narrow tunnel. His arms and hands felt strange and tingly. By the time his friends arrived, the young man was confused and disoriented. "He looked at me, and I could tell he didn't know who I was," another of the young men reported. They all nodded. "He's not like this," one of the friends said to the ER doctor. "He's never like this."

He had no medical problems except that he had been in a bicycle accident some months earlier, which had left him with a broken elbow and a dented helmet. The patient didn't smoke, didn't drink; nor did he use illegal drugs—at least as far as they knew. He was a slender man, fit and well kept. He wore jeans and a button-down shirt that was now damp with sweat and vomit. The physician smelled no alcohol on his breath, nor the fruity smell that would suggest he was a diabetic in need of insulin.

He didn't have a fever, and his blood pressure was normal. The meter on his finger showed that he was getting enough oxygen, although he was breathing deeply, as if catching his breath after a race. The rest of his physical exam was normal except that his fingers were bent, the muscles contracted as if he were gripping a ball.

The most striking abnormality was the young man's confusion. He could not follow even simple commands. He couldn't tell the doctor his name or where he lived. When asked, he guessed the year was 1990. (It was actually 2004.)

Dr. Lonny Shavelson, a physician with a couple of decades of experience in emergency departments, had frequently heard patients claim that they had been drugged, but he had found that this was rarely the case. Most drug use is voluntary. Still, the confusion and agitation were consistent with drug use. The doctor considered other less-common causes of confusion: an infection of the tough protective layer of the brain called the meninges would cause confusion—though usually there is fever or some other sign of illness.

Could the patient have become dangerously overheated at the sauna? The temperature taken orally was normal, but oral thermometers can be unreliable. A rectal temperature would be required to rule that out. On the other hand, his friends made clear that this man's illness became worse after he left the sauna, so the story wasn't really right for hyperthermia. Dehydration was possible, but it usually causes dizziness, not confusion. The opposite also seemed possible: water intoxication. More commonly seen in endurance athletes, it occurs when the sweating runner drinks too much water. The effort to prevent dehydration backfires, and the normal body chemistry becomes diluted. It is unusual even in runners, but still possible.

The physician turned his attention to another aspect of the patient's exam: he was hyperventilating. The young man was taking long deep breaths and had been since his arrival at the ER. Many of his symptoms were typical of hyperventilation: the tunnel vision, the tingling hands and feet, the tense, curved fingers. The question was, why was he hyperventilating? Could he have an injury affecting the breathing center of the brain? An injury that extensive would probably cause other neurologic effects, and none were apparent. You might also breathe this way if your blood became too acidic for any reason. Very deep breaths get rid of the carbon dioxide stored in your lungs and rapidly reduce the acidity of your blood. For example, this could happen to a diabetic in need of insulin. An overdose of aspirin can cause hyperventilation, confusion, nausea. He had had a recent injury—had he accidentally taken too many aspirin to treat the pain in his elbow? By far the most common cause of hyperventilation in the emergency room is anxiety. Still, it was rare for hyperventilation to produce this much confusion.

Could the hyperventilation be a reaction to the confusion and not the cause? The emergency room physician ordered an intravenous solution of saline to be started. If he had been at the sauna for several hours, he was possibly dehydrated. If he had water intoxication, the sodium would be helpful. He sent blood and urine to the lab to be tested for evidence of drug use—including aspirin.

The blood would also reveal an infection or an electrolyte imbalance. In addition, the doctor ordered a test of the patient's arterial blood to check its acidity. He briefly considered doing a spinal tap to look for evidence of an infection in the brain or a CT scan to look for a brain injury. These diagnoses, he thought, were less likely, and so he would order these tests only if the others brought no answers.

Then the patient's nurse took an oxygen mask, taped up the holes, and positioned it over the patient's mouth and nose. The usual treatment for hyperventilation is to have the patient breathe in his own exhaled breath to increase the level of inhaled carbon dioxide. He was also given a small dose of anxiety-relieving medicine.

The arterial blood, taken so painfully from the wrist, immediately confirmed the diagnosis of hyperventilation. The repeated deep breaths were making his blood too alkaline, and that was causing the visual changes and the clenched fingers. The rest of the results trickled in over the next hour or so. The drug screen was negative: no aspirin, no opiates, no ecstasy, PCP, or cocaine. No evidence of infection was found, either. However, his blood chemistries were quite abnormal: the sodium in his blood was dangerously low.

The brain is exquisitely sensitive to the exact right balance of sodium and water. When they become abnormal, nausea and confusion follow. Hyponatremia, or having low sodium, can cause seizures, coma, and even death if untreated. The IV of saline the patient was already getting would replace the electrolytes he had lost. "He felt as if he were drunk," the physician explained to the patient's friends. "And he was—he was drunk on water." The patient began to improve in the next hour. The three friends kept the doctor up to date on the patient's progress in figuring out the year. "He's up to 1999," they said gleefully. When he finally reached 2004, the group cheered. The patient was finally able to fill in some of the blanks. He had been worried about getting dehydrated at the sauna so he drank lots of water—lots and lots of water.

The patient was able to go home late that night, but it took him a week to really get back to normal. "It's amazing," one friend told me later. "To think that breathing, sweating, and drinking water—things we do every day—can do all that to you."

A River of Confusion

Dr. Jon McGhee, a second-year ER resident, casually greeted the patient and her fiancé as he entered the darkened hospital cubicle. "So, what's going on?" he asked the patient, who was also a doctor and a friend. The two had suffered through their intern year together—an experience that is the start of many enduring friendships.

She looked okay, the resident thought to himself, a little relieved. But her heart was racing—150 beats per minute. Her blood pressure was high, and she seemed anxious. But she didn't look sick. Then she began to speak. A wild river of words poured from her. Random words, meaningless sentences. There were snatches of sense scattered throughout, but they were drowned in the rushed torrent. McGhee looked at the young man, who nodded. This was why they had come.

The patient had been fine all day, the fiancé explained, but after dinner she started saying that she felt queasy and lightheaded. Within an hour, these symptoms began to worsen. She felt weak, she told him, sick and hot. Then she began crying uncontrollably, and when she spoke, she made no sense. That had really scared him.

The patient was twenty-seven, had an athletic build and no significant medical problems. The year before, she fainted a couple of times, but an extensive cardiac work-up hadn't revealed anything. She was taking an antidepressant, Paxil, and occasionally used another antidepressant, Elavil, to help her sleep. She didn't smoke, rarely drank, never used illicit drugs, and jogged daily.

When the doctor turned on the light to examine the patient, she cried out. The light had been bothering her since they got there, her fiancé told him. McGhee turned the lights down and began to examine the patient. She had no fever. Her mouth was dry, and her skin was quite warm though not sweaty. The rest of her exam was normal. An EKG showed no abnormalities beyond the rapid heart rate.

McGhee thought carefully about his friend. For almost anyone with a change in mental status, illicit drugs had to be on the list of possible causes, as unlikely as it seemed in this case. In addition, she took Elavil, which could cause many of these symptoms when taken in large quantities. Could she have taken an overdose? That could cause the rapid heart rate and confusion. But the most dangerous side effect of an Elavil overdose is dangerously low blood pressure. Hers was dangerously high. Perhaps the patient was bipolar and had moved from depression to mania? Or could it be something different? Could she have too much thyroid hormone? The thyroid is the flesh-and-blood version of a carburetor, regulating how hard the body's machinery works. Too little of this hormone, and the body slows down. Too much, and it speeds up.

He questioned the patient's fiancé. Had she shown signs

of mania? She had a history of insomnia, and sleeplessness was one sign of both mania and thyroid overload. How was she sleeping? Until this evening she had been fine, he insisted. She had been depressed, but not since starting the Paxil. And her sleeping was no worse than usual. He paused. There was one other thing: after dinner, he had felt a little funny, too. Not as sick as the patient, but his heart had been racing, and he had felt nauseated and jittery, though he felt fine now. They had eaten some lettuce from their garden that night. Could their symptoms be related to that? The resident immediately thought of a recent patient who had almost died from pesticide poisoning. He had been delirious like this patient, but had not had the elevated heart rate or blood pressure and had been sweating profusely. Still uncertain, the doctor ordered a few routine blood tests to look for the presence of an infection or an abnormality in her blood chemistry or thyroid hormone. He also ordered a urine test to look for illegal drugs and Elavil, the medication she used for sleep.

As he waited for the results, the patient became more agitated. She kept getting out of bed and walking into the chaotic hub of the emergency room, putting on gloves and picking up charts as if she were at work. At times, she seemed to hallucinate.

Over the course of the night, the patient's test results trickled in. The blood tests were normal. Her thyroid hormone wasn't too high. The drug screen was completely negative. What was going on?

By dawn, the patient's blood pressure came down. She was less confused. Her speech cleared. But she was still far from normal. Was this part of some underlying illness? Could it be linked to her earlier fainting spells? Was she having tiny strokes? Was she showering her lungs with little clots? Her symptoms weren't typical of any of these, but they didn't seem typical of anything else, either. Her cardiologist and a neurologist were consulted. She had an MRI of her brain to look for a stroke and a CT of her chest to look for tiny clots. Normal. After four days, the patient recovered, and she was discharged, her diagnosis still unknown.

At home, the patient worried about her brief descent into madness. That afternoon, she wandered out to the garden to do some weeding, and her attention was drawn to an uninvited guest growing there. Among the green and purple lettuces she had planted were several strikingly beautiful white and yellow flowers. Blossoms that hadn't been there before and that she was certain she had never sown. Before it had flowered, could this plant have been mistaken for lettuce and ended up in her salad? She pulled the plants up by their roots, put them in a baggie, and drove to a nearby nursery. As she pulled the plants from the bag to show the owner, the woman exclaimed: "Don't touch those! They're highly toxic. That's jimsonweed." Also called the devil's trumpet and sometimes locoweed, this plant has been known for centuries to cause a temporary kind of madness, the woman explained.

The symptoms caused by this class of plant are well known, and a mnemonic is taught in medical school to identify them: mad as a hatter, blind as a bat, dry as a bone, red as a beet. As it turned out, the patient had had all the classic symptoms. The plant's toxin affects the eyes, because it dilates the pupils and makes them very sensitive to light. She was also quite flushed, according to her fiancé, though both these symptoms were missed by McGhee, it seems, because he had turned down the light for his friend's comfort. He noted that her mouth and skin were dry and the madness was clear, but alone they weren't enough. By the time the other doctors saw her, most of these symptoms had resolved.

The doctor was being kind in keeping the lights low, but I think there's more here. McGhee didn't insist on being able to see—the way he would have insisted on drawing blood or getting the CT scan—because in this age of high-tech medicine, I think we no longer really believe that the physical exam can be an important diagnostic tool. Too often we simply go through the motions, never imagining that what we can observe will provide the kinds of answers that our machine-driven tests routinely do. Ultimately, that loss of faith can become a self-fulfilling prophecy.

In this case, the patient did fine without a diagnosis. She figured it out herself. I asked her later why she thought she had been so much more affected by the jimsonweed than her fiancé. "I'm not really sure," she answered. "Maybe I ate more of it. Or maybe it was the plant plus the Paxil. Antidepressants can cause some similar side effects." But it was a good lesson, and she hoped to publish her story as a case report in a medical journal.

The Sadness Signs

"Have you been to see your doctor?" the woman asked her seventy-two-year-old mother anxiously. Her mother had come from Miami to visit her in New York. They hadn't seen each other for a couple of months, and her daughter hardly recognized her. Her mother had been slender; now, she looked emaciated. Her usually bright eyes peered dully above her newly prominent cheekbones.

But it was more than that—the unrelentingly cheerful, energetic, outgoing woman she had known her entire life had disappeared. Now her mother spoke of nothing but how awful she felt and spent most of her day in bed.

It started a couple of months earlier when the mother and her partner were traveling in Italy. During their month-long vacation together, the woman began to feel irritable. She had fallen in love with this man eight years earlier—two years after the sudden death of her husband. And their years together had been happy ones. But on this trip, everything about him, about their relationship, began to grate on her. Suddenly she didn't want to travel with him; she didn't even want to see him. Indeed, she didn't want to see anyone.

When she got home, she felt no better. She was a psychologist and recognized the symptoms of anxiety. She had never felt this before, but she had seen it in her patients. She went back to the psychiatrist she'd seen a few times after her husband died. Yes, the therapist agreed, she did seem anxious. Also depressed. The woman accepted the psychiatric diagnoses but told her therapist that it wasn't just her mind; her body felt like it was too fatigued to do the work of living. But of course, the therapist told her, your mind is part of your body. People, especially older people, often feel symptoms of depression in their bodies—feeling sick and tired rather than sad.

The woman started taking an antidepressant and saw the psychiatrist once a week. When that didn't help, the psychiatrist tried another drug. When she was still no better, she saw another psychiatrist, who added an antipsychotic. By the time the mother went to visit her daughter, she was on four medications: one for anxiety, two for depression, and one for her insomnia. Despite all that she remained anxious, depressed, and unable to sleep.

Her daughter worried. Why wasn't she getting better? "Go see Cindy," her daughter said—Dr. Cindy Mitch-Gomez, her mother's longtime physician.

Once she was back home in Miami, the woman went to her doctor. When Dr. Mitch-Gomez saw her patient, she, too, was alarmed. She had lost weight. And she seemed to have lost so much more; she slouched in her chair as if it were too much work to sit up straight. During her routine physical five months earlier, the woman was her normal, lively self. Now she was a skinny, sullen shadow.

The patient explained to Dr. Mitch-Gomez that she had suddenly developed anxiety and depression. She was taking four medications and meeting with a therapist, and still she felt terrible. She gave up her daily morning exercise class because she didn't want to see anyone. She wasn't suicidal, but she couldn't bear feeling like this for the rest of her life.

She was one of the last patients of the day, so Dr. Mitch-Gomez settled in for a full investigation. The patient had a few vague complaints—she felt nauseated sometimes and occasionally awoke covered in sweat, as if she were going through menopause again. To Dr. Mitch-Gomez, it seemed clear that something beyond depression had to be going on. Although the patient was focused on the psychiatric symptoms, her doctor was worried about the fatigue, nausea, weight loss, and sweat. The patient had been treated for breast cancer nearly fifteen years earlier. Could it have come back and invaded her liver, lungs, or brain? It would be unusual after so many years, but not impossible.

If not cancer, what? Hypothyroidism is common in older adults. So is vitamin B_{12} deficiency. Each can cause depression. She frequently visited the Northeast; could this be some form of advanced Lyme disease? She sent the patient to the lab for blood tests and ordered a chest X-ray, a scan of her brain, and another of her abdomen and pelvis.

The blood tests came back quickly. Her thyroid was normal; so was her vitamin B_{12}. It wasn't Lyme. The chest X-ray showed nothing. The head CT was completely normal. But the scan of the pelvis was not. There was a small abnormality around the left ovary and uterus. A transvaginal ultrasound revealed a small mass on the ovary. After some testing, a gynecologist recommended removing the ovary and uterus.

The patient's partner heard the news before she was even

out of the OR: It was ovarian cancer, which had spread to a fallopian tube. But the surgeon assured him that they had gotten it all.

After the surgery, the patient no longer had cancer, but she was still depressed. Dr. Mitch-Gomez referred her to a psychiatrist who specialized in cancer and depression. She wasn't sure how the two were related—but it would be an odd coincidence, she thought, if they weren't.

The patient went to see Dr. M. Beatriz Currier, an expert in the biochemical connection between cancer and depression. Patients with cancer are up to three times as likely to have depression as those without cancer, Dr. Currier told her new patient. It's not simply that having cancer is depressing. It is that some cancers—perhaps most of them—can trigger the body to release chemicals that signal the brain to develop depressive symptoms. It's a phenomenon first reported in 1931, when Dr. Joseph Yaskin, a neurologist at the University of Pennsylvania, published a case series on four otherwise-healthy middle-aged patients who were initially thought to have late-onset depression and anxiety and who within months were found to have pancreatic cancer. The depression, Yaskin hypothesized, was a reaction "on the part of the central nervous system to the toxic or metabolic changes produced by the progressive visceral condition"—the cancer.

More recent research suggests that the body's response to an injury like that caused by a tumor or an infection is to release messenger chemicals known as cytokines. These messengers communicate with other parts of the body—the immune system, the brain, the gut—to trigger a response to that injury. Different messengers elicit different responses, and the cytokines triggered by certain cancers have been found to set off not only a robust immune inflammatory response but also the neurological changes that cause depression. Some researchers hypothesize that the behavioral changes caused by depression and anxiety—lethargy and avoiding contact with others—might have provided a survival benefit in the face of infection or injury.

Your body released a flood of cytokines in response to the invasion of abnormal cells, Dr. Currier told her, and these chemicals triggered your brain to become depressed. For the first time in months, the patient felt a flicker of hope. Hearing that she was depressed for a reason somehow made the depression a little easier to bear. With the cancer gone, she asked Dr. Currier, would her depression subside? That's the theory, Dr. Currier replied.

And it did, slowly. Over the course of a year, Dr. Currier weaned the patient off the psychiatric medications she was taking. She gained back her lost weight. She started going to her exercise class again. "I'm back," she told me triumphantly—and it was all thanks to her doctor, who had suspected that what looked like a familiar depression could be something more.

A Terrible Madness

"This is all a big mistake." The man sat leaning forward in the flimsy plastic chair. His eyes were bright, and his arms were wrapped around his torso, hands gripped tightly to his thin shoulders. "Do you know who I am?" he asked. "Well, do you?" He paused, then added, "I need to call my lawyer." His face contorted into a grim version of a smile, then he began to pace the small room. He was lanky, with broad shoulders, but his clothes were dirty and hung from his bony frame as if he'd lost a lot of weight since they were new. Jessica McCoy, a third-year medical student, glanced nervously at the doctor who had come with her to interview the patient. He nodded his encouragement, and she turned back to the patient. "Tell me what brought you to the hospital," she asked once more. He'd complained of back pain in the emergency room, according to his slender chart. He had enemies, he explained to those doctors—enemies who had broken into his home and injected him with poison. That was what was causing his back pain.

"I told the other doctors, and now they won't let me leave," the patient said. "This is an outrage. I am the richest man in the world. I need to call my lawyer." As he spoke, he gestured wildly, and his face distorted at times into a strange, involuntary smile—at odds with the intensity of his refusal. But McCoy was patient, and slowly the story emerged. The back pain was bothering him today, but he thought maybe he'd had it for a while. He hadn't been able to sleep for several days and hadn't been able to eat for even longer, and he didn't know why. He was thirty-eight, he didn't care much for doctors and he had never seen a psychiatrist; he had never been in a psychiatric ward—until now. He was a famous singer, he told them. He'd cut albums, gone platinum, toured the globe. Why hadn't they heard of him? He didn't smoke, he rarely drank, and he had never used drugs. He had no family. He spoke rapidly—his words crowding one another, making them unintelligible at times. Occasionally he turned his answers into rhymes, rapping bits of his story.

It was clear to both McCoy and the doctor, Matthew Hurford, a second-year psychiatry resident, that they would not be able to make a diagnosis based on this history alone. They needed to examine the patient; they needed blood for tests. The man was clearly manic. His energy was frantic; he couldn't eat or sleep. Words poured out of his mouth like water from a high-pressure hose. But what was causing it? Drugs—crack or methamphetamines—were probably the most common cause of such mania, but the patient denied any drug use. Abnormalities in his body's chemistry—too much thyroid hormone, too little sodium—can also change the way the brain works. Or was this really a psychiatric disease? Could it be the first manifestation of bipolar disease—the manic phase of manic-depression? He was a little old for that. Bipolar disease and schizophrenia typically appear in late adolescence or early adulthood and often run in families. Or was it a disease of the brain? Organic brain disease can mimic mental illness but often reveals itself in characteristic findings on physical examination. Despite McCoy's encouragement, the patient was adamant: they could not examine him; they could not take any blood. "I know my rights," he said. "No blood, no blood." There was nothing wrong with him, he insisted. Then he sat back down in the chair, wrapped his arms tightly around his chest, and refused to speak again.

The student and the doctor left the locked ward and tried to put the story together. Drugs seemed unlikely; the ER had been able to send some urine to be screened for the most common drugs—all tests were negative. Though he denied any family history of mental illness, it wasn't clear how much of this history could be trusted. What they needed was more information. In the ER, the patient had given the name of a woman as an emergency contact. McCoy returned to the patient and asked if she could call the woman. "Sure," he said. "She'll tell you who I am. And then you'll have to let me out."

McCoy called the woman. "Thank God he's all right," the woman said, clearly relieved. The patient had been missing for days, and one of his sisters had filed a missing-person report with the police. The woman had known the patient for two years and noticed that he had become increasingly withdrawn and quiet—and strange. He'd stare for hours at the television with the sound muted. He seemed paranoid and suspicious. "I still love him, but he's like a different person now," she told McCoy. The woman confirmed parts of the patient's story: He'd never been to a psychiatrist; he didn't smoke, drink, or use drugs. He loved to play music. He worked as a cook in a nursing home but recently lost that job because of this strange behavior. Both of his parents were dead. He still had family, however: an eighteen-year-old son, now in college, a brother, and two sisters. "His mother died young of some rare genetic disease," the woman said. "I don't know what it was." She paused. "You know, I've been wondering if he's got it, too." McCoy hurried to find Hurford and tell him this news.

There are a number of rare, hereditary diseases that progress slowly and can manifest as psychiatric illnesses. Wilson's disease, caused by an overload of dietary copper, can produce tics and irritability. Acute intermittent porphyria can also cause psychosis, but severe abdominal pain almost always heralds its onset. The doctor, however, immediately focused on Huntington's disease. This neurological disease, which can cause mental illness symptoms (usually depression), is accompanied by a movement disorder known as chorea, derived from the Greek word for "dance." The grimaces and dramatic gestures seen in this patient were so typical of the disease that it used to be called Huntington's chorea. Each child of an affected parent has a 50 percent chance of getting the disease.

So once again, Jessica McCoy headed back to the patient's room, accompanied by Dr. Hurford. When they asked the patient about his mother's illness and death, his answer was quick and definitive. Yes, she had Huntington's, but he was certain that he did not. He hadn't been tested, didn't want to be tested, didn't need to be tested, because he didn't have it. That evening, McCoy called the patient's older sister. She confirmed that their mother died of Huntington's. The oldest brother had it as well and was now in a nursing home. It saddened her to hear that her younger brother probably had it, though she had begun to suspect as much when she heard of his odd behavior. "Now I guess I have to worry about his son, too."

It took several days—and the help of both sisters, the patient's son, and an assortment of nieces and nephews—to persuade the man to let the doctors take the blood needed to confirm the diagnosis. They had already gotten him to agree to take antipsychotic drugs, and soon the paranoia and delusions began to subside. By the end of the week, he was discharged to the care of his family. The test came back positive a few weeks later.

I called the patient's older sister—the matriarch of the family—to find out how he was doing eighteen months after the diagnosis. The medicines had quickly made him almost normal, she reported, but even then he couldn't believe he had Huntington's. He stopped taking them soon after, as if he preferred his delusions to the reality of living with Huntington's disease. He was staying in a local shelter. Family members saw him occasionally, but he always refused to come home. Maybe running away from the family was his way of running from the disease itself. "I understand that," she told me. "And what can I tell him? Will our love change what's going to happen to him? He knows it won't. All we can do is care, and we'll do that no matter where he is."

High-Pressure Crazy

The patient lay on the bed, her eyes wide with fear as she struggled for breath. The nurse at the bedside looked almost as scared. She turned as Dr. Kennedy Cosgrove entered the hospital room and said, "I can't get a blood pressure, doctor—her pressure is too high for me to measure." Cosgrove felt his own blood pressure soar. Most patients in this psychiatric ward of Stevens Hospital in Edmonds, Washington, were physically healthy, and Cosgrove, a psychiatrist, hadn't managed this type of emergency since his internship. He ordered an EKG and quickly phoned the internal-medicine doctor on call.

Ten days earlier, the patient was brought to the hospital's emergency room by the police. According to their report, she phoned her teenage son to say goodbye—she was going to take her life. He and the police found her at home, shouting, incoherent, weeping.

When Cosgrove met her later that day, his first thought was that despite her erratic behavior—which wasn't unusual in this ward—she looked different from his other patients. Her hair was well cut. Her nails were clean and manicured. She looked tired and disheveled, but she didn't look chronically mentally ill.

After introducing himself, Cosgrove asked the patient if she knew why she was there. Tears filled her eyes. She couldn't take the disappointment of life anymore, she told him. He nodded sympathetically. She shifted restlessly on the bed. "I've had seven death attempts on me—by the police!" she shouted, suddenly angry. Her eyes narrowed suspiciously. "Have you heard this?" There was a conspiracy against her—organized by the state of Washington and the Boeing Company. Sometimes she could even hear them talking to her—their voices coming from inside her own brain. She laughed giddily and then became angry again. "Get out! Get out! Get out!"

In the nursing station Cosgrove reviewed the information collected on the patient in the emergency room. She was thirty-nine. Divorced. Lived alone. She took two medications for high blood pressure, as well as an antidepressant and a stimulant, Concerta, a long-acting form of Ritalin, for a diagnosis of attention-deficit disorder.

It wasn't clear how long she had been on any of these medications, but Cosgrove quickly focused on the stimulant. Psychosis and mania were rare but well-documented side effects of Concerta. Was that the cause of her symptoms or was this simply a manic episode in an underlying bipolar disorder?

Cosgrove stopped the stimulant and tried to start the patient on antipsychotic and mood-stabilizing medications. She refused to take them, so he gave the medication by injection. Slowly her behavior began to change. The wild swings of emotion and outbursts of anger became less frequent. But strangely, her disordered thinking and paranoid delusions persisted. Usually these symptoms improved and worsened together. And now she had this significant spike in blood pressure. Were the high blood pressure and the psychosis linked?

Dr. Michelle Gordon, the internist on call, hurried to the patient's bed. She took her blood pressure. It was critically high—240 over 110. She quickly transferred the patient to the intensive care unit.

Gordon ran through the short list of possible causes of this kind of rapid elevation of blood pressure in a young woman already on antihypertensive medications. By far the most common was illegal drugs. It seemed unlikely that she would have access to them in a locked ward, but Gordon would check. A narrowing of the arteries that deliver blood to the kidneys could also cause this kind of intermittent hypertension. While this was usually a disease of the elderly, sometimes, for reasons that are not well understood, it could be seen in younger women as well.

The third possibility on Gordon's list was a tumor that was secreting too much of one of the hormones that raise blood pressure. Most of these hormones are made in the adrenal glands—tiny organs that sit on top of the kidneys. One hormone, aldosterone, controls the amount of salt in the body. Too much salt causes blood pressure to rise—sometimes significantly. Or could the patient have a pheochromocytoma, a rare tumor that causes the adrenal gland to produce too much steroid hormone—not its namesake, adrenaline, but another hormone, cortisol. An excess of either of these adrenal hormones could send blood pressure soaring.

In the ICU, Gordon started the patient on an intravenous medication that brought her blood pressure back into the normal range. Then she ordered an ultrasound to assess the blood flow to the kidneys and measure the size of the adrenal glands. These tumors often cause the gland to enlarge. Gordon stood by the patient as the ultrasound technician ran the transducer over the slender woman's abdomen. He pointed out the blood flowing to the kidneys. It wasn't completely normal, but the arteries didn't appear narrow enough to affect the blood pressure. The fuzzy picture on the monitor shifted as the tech tried to get a good view of the kidney. "Will you look at that?" the tech exclaimed. The right kidney he pointed at looked normal, but the adrenal gland on top of it was hugely enlarged.

Gordon sent off samples of blood and urine to identify which of the key hormones was the culprit. She suspected this was a pheochromocytoma. Though these are very rare tumors and the patient had not complained of the headache, sweating, and rapid heart rate that are their hallmarks, the surges of the stimulating hormones seen in this disease could account for the spikes in blood pressure. When the results came back the following day they showed that Gordon's hunch was correct—she did have this rare steroid-producing tumor.

As soon as Cosgrove heard about his patient's diagnosis he began to wonder if her psychiatric symptoms could also be caused by the excess steroid hormone. He had taken her off Concerta, the stimulant, because he knew that it sometimes causes psychosis and mania.

If the stimulant drug could sometimes cause psychosis and mania, could this stimulant-producing tumor do the same thing? Cosgrove found several papers describing patients who, like this woman, had pheochromocytomas as well as mania and psychosis. The symptoms resolved once the tumor was removed. It was unusual, but it had been reported.

The patient was transferred to a larger hospital for the delicate operation to remove the tumor. When she returned to the psychiatric unit, she was still paranoid, still delusional. But Cosgrove was patient—in the papers he read, recovery took time. Over the next several weeks, the patient's thinking began to clear. Her paranoia ebbed. The mania disappeared. By the time she was discharged, a month after her operation, her blood pressure was normal, and so was she. Her doctors slowly peeled away the medications for her hypertension and mental illness. And finally, after a year and a half, she was medication-free and remained so three years after her surgery, when I spoke to her.

At that point it seemed clear that symptoms of this tumor had started several years before she ended up in that emergency room. But her symptoms were odd and intermittent—a sharp pain in her arms and in her chest, anxiety, transient high blood pressure, episodes when she was too jumpy to concentrate. The mania and delusions that ultimately led to her hospitalization and diagnosis began after she was diagnosed with attention-deficit disorder and started taking medication to treat it the year before. After that, her life became chaotic—her three children could no longer live with her. The double dose of stimulants—from her tumor and the drug—seemed to trigger her psychotic break.

The patient told me she could barely recognize the person she was before the tumor was removed. After she left the hospital, she sent Cosgrove a note: "Thank you for giving me my life back."

PART VI

Out Cold

Passed Out on a Saturday Night

The young woman lay on the stretcher, her eyes closed, her arms and legs in constant, restless motion. At the doorway, the patient's mother stood, unable to move, petrified by the image of her healthy, exuberant twenty-year-old daughter now pale and unresponsive. She went to the bedside and stroked her daughter's sweaty cheek. "What have you done, my little girl?" she whispered in the young woman's ear.

The young woman appeared not to hear, to not even know her mother was there in the emergency room of the Upstate University Hospital in Syracuse, New York. All that the ER doctors could tell her was that her daughter was dropped off early that morning by a young man. The triage nurse spoke to the man, who reported that the woman had gone to a concert with him the night before. They had been separated, and when they met up at the end of the evening, she had seemed happy—euphoric even. She slept on his couch, and that morning when he tried to wake her, she wouldn't even open her eyes. That was worrisome, but he let her sleep, he told the nurse. Later she became incontinent and began vomiting. That's when he finally took her to the emergency room. Then he left.

Waves of guilt and anger swept over the mother. She had waited up all night for her daughter. The girl always came home when she said she would, but the night before she hadn't. At six that morning, the mother drove to the apartment of the friends her daughter had gone out with. Yes, her daughter was there, a sleepy, hungover-looking young man told her. She was asleep. Worn out with worry and fear, the mother's relief turned to anger. How could her daughter have forgotten to call and let her mother worry all night? Tell her she's in big trouble, she told the young man, and then she left. Now she asked herself, why hadn't she gone into the house and tried to bring her home? How could she have left her there? And what kind of friend would bring a sick girl to the ER and then just disappear?

The young woman was unable to respond to her name, answer any questions, or follow even the simplest commands when she was examined by the ER doctor, Lauren Pipas. Only when the doctor rubbed her knuckles against the woman's chest—a maneuver used to check a patient's ability to respond to pain—did the woman seem to have any awareness of the world around her. She moved, trying to elude the knuckle pressure, and moaned, but even then didn't open her eyes. She had no fever, and her heart was beating normally.

In a patient this age who comes in unresponsive and with no obvious illness, drug intoxication or overdose is the most likely cause. Pipas ordered a urine drug screen, which would pick up the most commonly abused drugs. She also tested her blood for acetaminophen and salicylates, the active ingredient in aspirin. An overdose of either of these common over-the-counter medicines can kill if not diagnosed and treated quickly. And though an overdose seemed most likely, Pipas needed to make certain there was nothing else going on. The urinary incontinence reported by the young man suggested a possible seizure. So she also ordered an EKG, a chest X-ray, a head CT, and routine labs, including a blood count and chemistry test. She would also test for thyroid disease and pregnancy, both common in young women who came to the emergency room.

The test results came back quickly. The EKG and CT were normal. She had none of the most commonly abused drugs in her system; there was no alcohol, no marijuana, no opiates. The urine test was positive for amphetamines, but that seemed an unlikely cause of her unresponsiveness. Her serum sodium—a key blood chemical that is central to most body functions—was dangerously low. Low sodium, or hyponatremia, can cause a loss of consciousness and seizures. If not corrected, it can cause permanent brain damage, even death. This remarkably low level of sodium would clearly explain why the young woman was unconscious. But why was her sodium so very low? It was an important question but one that would have to wait until the woman was stabilized.

Pipas ordered a sodium solution to replace the essential electrolytes. She gave the woman a sedative for the agitation. Then she called the team from the intensive care unit. Shaun Cole, a fourth-year medical student rotating through the ICU, was the first to arrive. He reviewed the woman's chart and eyeballed the patient. The sedative had stopped her restless movements, but she was still responsive only to pain. He asked her parents and her brother, who had also come to the hospital, to tell him about the woman in the bed, and particularly about the past twenty-four hours.

She had been out with some friends, the family told him. She was a good student, and she worked weekends in a local restaurant to help pay her way. That didn't leave her a lot of free time. Had she used drugs before? Cole asked. She drank alcohol but beyond that, nothing, the parents assured him.

What about her cellphone? The brother rummaged through a bag of his sister's belongings and pulled out the phone. She had made a bunch of phone calls to her friends—no surprises there. Then Cole started clicking through the instant messages. He saw several references to "Molly"—who was Molly? Google quickly provided the answer; it was another name for Ecstasy, the amphetamine-derived drug often used at concerts and raves. But how was that drug linked to the woman's current condition?

That afternoon Cole looked for the link. Ecstasy has been shown to cause this kind of extreme hyponatremia, especially in young women. The drug affects the brain and kidneys in ways that promote water retention, which dilutes the sodium in the body. It wasn't clear why an individual would have this kind of reaction, but case reports from emergency room visits indicated that it was not linked to higher doses and could occur in those who had used the drug before. It is a dangerous side effect. Nearly one in five who developed hyponatremia because of the drug died. Others had permanent brain damage.

When the patient finally woke up a week later, it was clear her brain was injured. Her speech was jumbled and slow. Her vision was impaired. She had to learn to read and write again. She spent the next several months working to regain all she had lost. Incredibly, she graduated from college just one semester behind her classmates. She told me that she had rarely done drugs—she'd taken Ecstasy only once before. Now all that remains of the terrible experience are the memories of how hard she had to struggle to get herself back to who she had been.

As the young woman worked to recover, her mother wanted to prevent what happened to her daughter from happening to anyone else. The young woman had been dangerously ill for some time—possibly hours before anyone took her to the ER. Why? Were her friends afraid that they might be arrested if they revealed that they used drugs the night before? Had they not worried about getting in trouble, perhaps they would have taken her earlier.

They had acted foolishly, but the mother could understand their fear. She decided to do something about it. With the help of her state senator and others, the mother lobbied for the passage of the Good Samaritan law in New York State. This law, which was pioneered in New Mexico, protects those who seek medical help for someone with a drug or alcohol overdose from being prosecuted. The law was passed in 2011.

Frequent Fainting

The middle-aged woman perched on the edge of a plastic chair as the doctor explained his thoughts on why her son was having persistent headaches. Suddenly, she toppled forward, collapsing onto the linoleum floor. Dr. Philip Ledereich hurried over to the woman. "Call 911," he shouted to his nurse. "The patient's mother has fainted."

Ledereich, an ear, nose, and throat specialist in Clifton, New Jersey, first met the mother a couple of weeks before, when she herself came in as a patient. She was fainting several times a day, and no one knew why. Ledereich hadn't been able to figure it out, either. Despite that, she brought her son to see him for the treatment of a chronic sinus infection. Ledereich was describing various treatment alternatives when the woman pitched to the floor.

She had been having these spells almost daily for the past several months, she told him at their first appointment. She was forty-nine and a nurse, and had considered herself quite healthy until one Saturday nearly three months earlier. That day she had just put on her shoes to go to a bar mitzvah, and as she straightened up she felt a fluttering sensation in her stomach. The next minute she was on the floor. Her husband rushed to her side. She could hear him calling her name, but she couldn't answer him; she couldn't even open her eyes.

And then, just as quickly as it started, it was over. She felt just fine. She didn't want to go to the hospital, she told her husband. She wanted to go to the synagogue. And so they did, walking a mile. At the coffee hour following the service she started to feel that fluttery sensation in her stomach. Was she going to faint again? She was almost at the door when she collapsed. Eventually, her husband persuaded her to go to the hospital.

She spent two nights in the cardiac-care unit as doctors looked for any of the irregular heart rhythms that could start as a fainting spell but might end in death. They found nothing. She had a head CT scan and lots of blood tests. Everything was normal, so she went home.

Syncope—the medical term for fainting—is common. Up to half of the population will faint at least once in their lifetimes. Most of the time the cause is benign and transient. The trick for doctors is to identify those cases that are neither. When a heart beats too rapidly, too slowly, or too erratically to deliver enough blood to the brain, you faint. If a normal rhythm isn't restored in time, you might never wake up.

Far more often syncope is triggered by dehydration or other causes of sudden low blood pressure. The best way to distinguish among these nonfatal varieties is to witness an attack. And so, before the patient left the hospital, she had a tilt-table test—a study designed to provoke a fainting spell. The patient was hooked up to blood-pressure and cardiac monitors, and strapped to a table, which is then positioned almost vertically. And then they wait. She stayed in that position for nearly an hour. A test is considered successful when the patient passes out and the monitors capture the cause. But the patient didn't faint. She went home hoping that whatever caused the two episodes had simply gone away.

But the next day she was driving to work and began to feel that now-familiar flutter in her stomach. She pulled off the highway just in time. When she awoke, she called her husband, who took her directly to her doctor's office. Her internist was as baffled as the doctors she'd seen in the hospital. He sent her to specialists. One thought that these spells might be seizures rather than syncope. But a normal electroencephalogram (EEG) suggested otherwise. A highly recommended neurologist in New York carefully examined her and her now-thick chart and pronounced definitively that there was nothing wrong with her and that she should try to relax and maybe take up yoga.

That's when she scheduled the appointment with Dr. Ledereich. She thought she might have an inner-ear problem, and he had also been recommended by her friends. At that first encounter, Ledereich was not optimistic. He knew she'd seen many specialists. But he listened to her story and examined her. Like the doctors before him, he found nothing. She felt a little tired and had a little asthma, but other than these strange, repetitive spells, she was fine. He would get her records and then have her return. Meanwhile, when her son needed to see an ENT, she took him to Ledereich—and now she lay motionless on the floor.

"Don't call 911," the patient called out. She opened her eyes. "This happens to me all the time. I'm fine. Really."

Ledereich watched as the patient calmly sat up. "I know what you've got!" he told her excitedly. Her sudden collapse looked as if a switch had been thrown and all her muscles just turned off. Ledereich realized that although it looked like syncope, it wasn't; she hadn't actually lost consciousness. What she probably had, Ledereich told her, was something called cataplexy, and that meant that she also had narcolepsy. With narcolepsy, elements of sleep invade your waking hours, and elements of wakefulness intrude on sleep, leading to insomnia at night and persistent sleepiness during the day. Most patients with narcolepsy also have cataplexy. In this disorder, the total loss of voluntary muscle control that keeps us from acting out our dreams as we sleep interrupts our waking life, causing the sudden, dramatic loss of strength he observed. For reasons that are not yet known, these attacks are usually triggered by strong emotions.

"I was telling her about possible treatments for her son's chronic sinusitis," Ledereich explained to me, "and I said that if all else fails, we could try surgery. As soon as the word was out of my mouth, she hit the floor." For the doctor, the combination of three factors—the patient collapsing after experiencing stress (hearing that her son might need surgery), the fact that she could hear him ask the nurse to call 911 (indicating that she was not unconscious), and her rapid recovery—added up to a diagnosis of cataplexy.

The biology of narcolepsy is only beginning to be understood. Cells that make the proteins that keep sleep, and in particular REM sleep, at bay are somehow destroyed, and that allows bits of REM sleep to penetrate our waking hours. No one knows just what causes the destruction of these cells.

Recent research has shown that this is an inherited disorder. However, most of those who inherit the genes don't have narcolepsy. How this all works is still not well understood.

The patient hadn't complained of sleep problems, but when Ledereich began probing, they were there. Most of her adult life she had slept in two- to three-hour naps. Hearing this, the doctor was certain she had narcolepsy and cataplexy. A sleep study confirmed his diagnosis.

Treating cataplexy is difficult. Many patients end up on a drug sold under the trade name Xyrem, known on the street as GHB, the date-rape drug. It is a powerful, fast-acting sedative and helps those with cataplexy get the sleep they so desperately need. Her doctor warned her that it's only moderately effective, so the patient was thrilled when the spells stopped once she began taking it. But for reasons that neither the patient nor her doctors understand, after about six weeks, they returned. At first, just occasionally. Then almost daily.

The patient has learned to cope with her unusual condition; she no longer drives. And when she feels the warning signs, she tries to alert those around her to tell them not to worry. She's part of a small community, and by now, most know her well enough not to call 911. Most, but not all. Recently, at her son's bar mitzvah, she began to have that familiar fluttering in her stomach. She tried to warn the woman next to her, she said, but ran out of time. The woman shouted for help before others came over to quietly explain that this was an ordinary event with her. The patient paused and then told me with a quiet laugh, "I guess she doesn't get to synagogue that much."

Cold Case

I think I'm going to faint," the tall thirty-five-year-old man said, grabbing the handrail of the stairs that led up from the beach. His brother turned back to see his older sibling collapse onto the sandy wooden planking.

Lying on the walkway, the well-tanned man looked strangely pale; his lips were tinged with blue—although the ocean from which they had just emerged was very warm. And his hands were bright red and seemed somehow larger than normal.

The patient was awake by the time he arrived in the ER. His wife noted that some of the color was starting to return to his face. A doctor bustled in and began asking questions. The patient said he was fine until he went swimming that day. But after being in the water for a while, he began to feel lightheaded, and the skin on his hands and feet felt tight, as if it had shrunk in the wash. He made his way up to the beach and sat with his legs pulled up tight against his chest. He felt cold, he remembered, and couldn't stop shivering. His hands were so swollen by then that he had to take off his wedding ring, his wife added. He was no better by the time his brother came out of the water a few minutes later, and so they started back to the house they had rented with their families for the holiday. His heart was beating hard and fast, and his vision seemed to close in. Then he fainted.

He had no other symptoms. He didn't remember getting stung by anything while he was in the water. He had no medical problems—in fact he just had a checkup before leaving for vacation.

In the ER, the patient's heart was beating faster than normal and his pulse was faint. His hands and feet were red and swollen. All of his symptoms resolved over the next couple of hours. None of the tests suggested any underlying heart disease, but given the patient's age and symptoms, the doctor suggested that he follow up with a cardiologist when he returned from vacation.

Back at home, the patient did see a cardiologist. An EKG and a stress test were normal. The cardiologist couldn't determine what happened that day at the beach but didn't think it was the patient's heart. Relieved, the patient forgot about it—until it happened again. The next winter he and his family went to Peru for a little sunshine. He was body surfing when suddenly he felt that same lightheadedness. This time he was frightened.

He realized that if he fainted in the choppy surf, he could die. Emerging from the water, he sat down in the warmth of the January sun, waiting for the dizziness to pass. His heart beat hard and fast, and his hands and feet were red and swollen just as they were that summer day. This time he didn't faint, but he spent the rest of his vacation admiring the waves from the safety of the beach.

Back in the New York City winter, the patient noticed that his hands became red, swollen, and painful when exposed to the cold air. The patient's sister, a nurse, suggested that this might be Raynaud's phenomenon. Raynaud's is an exaggerated response to cold that causes fingers (and sometimes ears, nose, face, and toes) to change color when exposed to cold. Affected body parts turn white or blue as the blood vessels constrict significantly in response to cold and then turn red as blood flows back in.

While usually benign (although sometimes painful), Raynaud's can indicate the presence of serious disease. His sister urged him to see a rheumatologist and find out whether he had Raynaud's phenomenon. That's what took him to the office of Dr. Efstathia Chiopelas, a rheumatologist at New York University. The patient described the strange swelling and redness that affected his hands that winter and the two episodes of lightheadedness that came after swimming.

His exam was completely normal, but Raynaud's—if that's what he had—was most commonly triggered by cold, and the office was quite warm. The doctor disappeared from the exam room and returned with a basin of water and ice. She took the patient's right hand and plunged it into the freezing mix. If this was Raynaud's, his fingers should respond by turning white or blue. The change was immediate. The hand she placed in the water darkened to a deep red. And it was so swollen it nearly dwarfed its partner.

His reaction wasn't typical for Raynaud's, but Chiopelas knew autoimmune diseases can be varied in their manifestations. She would test him for some of the diseases linked to Raynaud's—just to make certain she wasn't missing anything—but she suspected that he had come to the wrong kind of specialist. His swollen hand looked like an allergic reaction known as angioedema. This severe localized swelling was sometimes an isolated finding but could also herald a severe allergic reaction—even anaphylaxis. "I don't know what you're allergic to, but it seems to be triggered by the cold," the doctor said. "I think you need to see an allergist."

The patient's wife was really worried. An allergy induced by the cold? She sat down at the computer and looked up "allergy induced by the cold." Up came something she had never heard of: cold-induced urticaria—an allergy to cold itself. As unlikely as it seemed that someone could be allergic to the weather, the description completely fit her husband's symptoms.

She made an appointment for her husband with Dr. Christine Fusillo, an allergy specialist. After hearing his symptoms and his response to the ice-water test performed by Chiopelas, Fusillo nodded. Their home diagnosis was correct. The patient was actually allergic to cold.

An allergic reaction occurs when some exposure triggers special white blood cells, known as mast cells, to release chemicals (including histamine—which is why antihistamines are used in treatment) into the bloodstream. These chemicals cause blood vessels to leak fluid into the surrounding tissue, resulting in the localized swelling and itching seen in many allergic reactions: the itchy, swollen eyes and red, runny nose of seasonal allergies as well as the constricting throat and shock of anaphylaxis. For most allergy sufferers, the trigger is a substance, but the environment itself can cause this kind of chemical release for a small subset of people, and among those, an allergy to cold is the most common trigger.

No one knows why some people develop this allergy, but young adults are most commonly affected. The allergist wrote out a prescription for a daily antihistamine and an EpiPen, a shot of epinephrine to counteract severe allergic reactions. He obviously should avoid the cold as much as he could, she told him, but the time of greatest danger to these patients isn't necessarily the winter, when it's easy to remember to dress warmly. Paradoxically, it is the warmer months that carry the greatest threat; that's when the sparkling waters of the ocean or neighborhood pool beckon or an unseasonably cool evening can catch you unaware and unprepared. It's not unusual that his worst symptoms came while swimming, when his entire body was exposed to the cold, the doctor said.

The patient told me that he had learned to live with his odd allergy since getting the diagnosis. The hardest part is explaining his condition to others. Recently, he went to pick up his daughter from her elementary school on a cold day. He was early so he entered the building to stay warm, although parents were supposed to stay outside. When asked why he came in, he explained that he was allergic to the cold. A teacher nodded and laughed. "Aren't we all?" she said as she walked off.

The Deepest Sleep

"I can't wake my wife up!" The voice on the phone was panicked. The couple was lying in bed when the forty-three-year-old woman began to snore, something she'd never done before. Her husband tried to wake her, but she didn't respond. He shouted her name; he shook her shoulder. Nothing. Terrified, he called 911.

He couldn't imagine what was wrong with his healthy, active wife, he told the medics when they arrived. That day had been pretty normal. His wife got home from work in the late afternoon. She made dinner and then went to her kickboxing class. She got home, put the kids to bed, and had a drink—vodka and cranberry juice. Then the couple went to bed. Normally the husband stayed up later than his wife, but this evening they turned in at the same time. As they lay in bed, talking, his wife began to snore. It was so sudden and unexpected that, at first, he thought it was a joke.

The medics tried to wake her. They called her name;

they shook her. They gave her Narcan, a drug that counteracts narcotics, even though her husband told them she didn't use drugs. All they got from her was a moan. They loaded her into the ambulance and hurried her to the University of Michigan Medical Center emergency room.

Dr. Robert Silbergleit, the doctor on duty at the ER that night, met the ambulance on arrival. "Comatose forty-three-year-old woman found by husband," the EMTs reported as they moved the patient into the area reserved for the critically ill. "Snoring. Unresponsive except to pain."

The patient, a thin woman, seemed to Silbergleit to be perfectly fit and healthy, except that she was asleep. He rubbed her chest hard with his knuckle—a sternal rub, it's called—a very painful maneuver, designed to elicit a response. "Stop," she moaned, but she didn't open her eyes. And she didn't wake up.

Silbergleit organized the possible causes of the woman's strange and unexpected somnolence.

Drug overdose: a common cause of a sudden loss of consciousness in an otherwise-healthy adult. Her husband said she didn't use drugs and the Narcan didn't do any good, so narcotics were unlikely. Nevertheless, Silbergleit sent off blood and urine to look for other sedating drugs or alcohol.

Traumatic brain injury: Had she been hit in the head during her kickboxing class? A well-placed blow could cause bleeding inside the skull, resulting in a loss of consciousness hours later. She hadn't complained of a headache or mentioned an injury. Still, Silbergleit ordered a CT scan of the head. Untreated, bleeding around the brain can cause permanent injury or death.

Stroke: The sudden onset of symptoms sounded like a stroke, even though the symptom itself, a sudden loss of consciousness, did not. Still, stroke needed to be considered, because the damage caused by a stroke can be reduced or even prevented by medications that open the clogged artery and restore blood flow throughout the brain. But these drugs have to be given within four and a half hours of the onset of symptoms. And because these are powerful drugs that can cause life-threatening bleeding, a definitive diagnosis of a stroke is essential.

The drug tests were all negative. Her blood-alcohol level was consistent with the reported single drink after dinner. The CT scan of her head didn't show any evidence of bleeding or a stroke. Silbergleit ordered a CT angiogram—an image that outlines the arteries of the brain—to look for any obstruction that would suggest a stroke. It, too, was normal.

Silbergleit spoke with Lesli Skolarus, a neurologist with special training in strokes. It was late, and Skolarus was at home. Silbergleit described the case and explained that he was planning to get an EEG to see if the young woman was having continuing seizures.

Skolarus hurried to the hospital, arriving around 1:00 a.m. By the time she finished reviewing the ER test results, two hours had passed since the patient first fell asleep.

Like Silbergleit, Skolarus was struck by the sudden onset of symptoms. Was this a stroke? If it was, it was an odd one. Because of the way arteries course through the brain, any obstruction will cut off blood and oxygen to only one side of the brain. So the typical stroke causes weakness or paralysis on one side of the body, and the patient is usually wide awake. The part of the brain that keeps us awake is known as the reticular activating system, or RAS. The RAS is usually fed by one artery on each side of the brain, a redundancy that provides important protection if either of these vessels should be blocked. A small segment of the population, however, has only one vessel delivering blood to the RAS, and a well-positioned clot to this single vessel, called the artery of Percheron, could completely block blood flow to the RAS and cause unconsciousness. Could that be the case with this patient?

Skolarus had seen this kind of stroke once before, when she was in training. It took days to figure out why an older man had suddenly lost consciousness. By the time they discovered this rare stroke, the damage was permanent.

Skolarus looked at the clock. Three hours had passed since the woman's symptoms started. If she had a clot in the artery of Percheron, there was still time to use a clot buster to reopen the vessel before the injury became permanent. But first they would need an MRI for a closer look.

A half-hour later the scanner thumped and clanged as the patient's brain slowly came into view on the monitor. Skolarus watched as the skull, then the top of the brain and finally the midbrain appeared before her. And there it was—a bright spot indicating damage to the RAS from a blockage to the artery of Percheron. Skolarus called the ER to get the clot-busting drug ready.

The medication was started just moments before the four-and-a-half-hour deadline. As the medicine dripped in, Skolarus showed the husband the damaged region. Then she had to tell him the bad news: Even if the medicine worked, there was still a chance she would not wake up. And if she did, she would probably have residual damage. Chances were, she would never be the same.

Suddenly the husband heard his wife's voice. He and Skolarus rushed to the bedside. She looked a little scared, but her eyes were open and she was talking. She knew her name.

She knew his name. She knew the name of the current president. She was back.

The patient stayed in the hospital for the next several days. She felt fine, but the doctors needed to figure out why she had this stroke. A sonogram of the patient's heart provided the answer. It showed that there was an opening in the wall that separated the right side of the heart from the left. Normally blood comes into the right side of the heart, then passes through the lungs, where it is oxygenated, before going into the left side of the heart. From there it's pumped into the rest of the body. Because lungs also serve as a kind of filter, trapping clots and other particles in its tiny capillaries, the hole in this patient's heart allowed a tiny clot from somewhere in the body to travel through the heart and into the brain.

When I spoke to the patient, she told me she feels back to her old self and is amazed and grateful.

"There were so many ways I was lucky that night. If my husband had come to bed at his usual time, he might not have noticed anything. If the neurologist hadn't come in, if the MRI hadn't been available—" She stopped, unwilling to even talk about the possibility. "It's all a little scary."

Faint of Heart

"You need to come to the hospital right away." The voice on the phone was kind, the instructions clear. But to the fifty-nine-year-old woman on the other end of the line, the meaning seemed elusive. Was something wrong with her heart? Looking at their monitors, miles away, the doctors said it was beating abnormally, even dangerously. But that seemed so unlikely. Sure, she didn't feel great; for the past week she had felt sluggish and tired, as if she were about to come down with a cold. But her heart felt fine: no chest pain, no palpitations, nothing.

She packed an overnight bag—just in case—and had a friend drive her to MetroWest Medical Center in Framingham, Massachusetts. In the emergency department, she was hustled into a room and hooked to a heart monitor; IVs were placed, blood was drawn. A thick plastic wand went over her chest to "interrogate" her pacemaker and download the data it held. Doctors had implanted it just six weeks ago—back when all these problems started.

She remembered standing at her kitchen counter, a month and a half before, chopping scallions for dinner. Then, suddenly, she was on the floor. She didn't remember falling, or even feeling faint: One minute she was upright, and the next she wasn't. She felt fine. She simply picked herself up and finished making dinner.

She wouldn't have given it a second thought if it hadn't happened again the next day. That time she was driving. Luckily, she was moving slowly, making a left turn. Suddenly her car was up against the curb, and a woman was standing at her window asking if she was all right.

An ambulance took her to MetroWest, where a blood test found high levels of troponin, a protein released by injured heart cells. In the cardiac-catheterization lab, a cardiologist fed a narrow tube through a blood vessel and injected a dye to illuminate the arteries supplying her heart. When one of these becomes blocked, blood flow to part of the heart is cut off, resulting in a myocardial infarction, or heart attack—and that can cause erratic heartbeats and sudden blackouts. But the doctors found no such problem: Her arteries were perfect. It definitely wasn't a heart attack. While they were looking around, though, her heart suddenly stopped. Doctors immediately began chest compressions and gave her a large shot of epinephrine. Her eyelids fluttered open. We think you should have a pacemaker, a doctor told her, calmly. She got one right then.

It wasn't clear why her heart had these occasional arrhythmias, but the pacemaker would keep it beating while her doctors tried to figure things out. For the next six weeks, she saw her cardiologist regularly. Each time, someone would ask if she had experienced any chest pain or palpitations. She hadn't. But around week 5, she noticed that she couldn't walk as far or as fast as she used to. The cardiologist did an echocardiogram—an ultrasound of the heart—and told her it appeared to be working just fine.

Then she got the phone call.

At the medical center, the cardiac monitor showed a rapid procession of white spikes, like tall skinny soldiers dashing across the screen. Every now and then one was slightly out of step, which only emphasized the regularity of the others. A nurse injected a beta blocker into her IV, and the squadron of spikes slackened its pace.

A technician entered to get another "echo," applying a gel-covered probe to her chest and revealing fuzzy images of her rapidly snapping heart muscle. She was a veteran of nearly a dozen of these tests by now, but this time the tech said something she'd never heard before. "There's some muscle damage," he announced.

"We're going to send you to the Brigham," her cardiologist told her—Brigham and Women's Hospital, about twenty miles away in Boston. Alone in the ambulance, the woman's composure melted away. Her husband had left that morning to go skiing with their son in Jackson Hole, Wyoming. When she tried to call, his phone went immediately to voicemail. He was probably still in the air. *Call me*, she texted him, as tears ran down her face.

At the Brigham, the cardiologist Dr. Garrick Stewart was reviewing the patient's records before she arrived. Her heart was rapidly breaking down in every way. Its rhythm was abnormal—that's why she'd blacked out. The rate was abnormal, too; when they had called her at home, her heart was racing at more than twice the normal speed. And the echo showed that the muscle itself was damaged. Few diseases could do all that. She could have a form of myocarditis—an infection or inflammatory process that attacked the myocardium, or heart muscle. A virus was the most common perpetrator, but there were others, too. An autoimmune disorder, like giant cell myocarditis, could cause the same kinds of widespread damage, but those were rare. Or it could have been an infiltrative disease like sarcoidosis, which doesn't actually destroy heart muscle but inserts abnormal cells into it, interfering with the way it does its work.

Of all these options, the most deadly was giant cell myocarditis. If that was what she had, she would need to be given medications to shut down her immune system immediately: The damage done by this disease is rapidly progressive and cannot be undone. Immediate, aggressive treatment is essential to prevent it from worsening. It was the rarest of the three rare disorders, but it was also the one the doctors could not afford to miss. She would need a biopsy of the heart to make that diagnosis.

When the patient arrived in Boston, there was more bad news. After the biopsy, she'd need an MRI. Her pacemaker wasn't a problem for the machine, but her wedding ring was. She hadn't taken it off more than a few times in the past thirty-five years, and now she couldn't. It had to be cut off. She felt strangely naked without it. And so very lonely. It was comforting to know that her husband was trying to get home, but it wasn't the same as having him there.

The biopsy results came back the next day. It was giant cell myocarditis (GCM), the most dangerous of the possibilities. This mysterious and deadly autoimmune disease has only been diagnosable before autopsy since the development of the heart biopsy some fifty years ago. Effective treatment has only recently emerged.

Stewart went to talk with the patient. The picture, he knew, was grim. Without treatment, most patients with GCM either died or had a heart transplant within months of diagnosis.

I have good news and bad news, he said. The bad news is that you have this very tough disease. The good news is that we know how to treat it. She would need several immune-suppressing medications, and she would have to undergo extensive evaluation to see if she qualified for a heart transplant, in case one was needed. With aggressive treatment, a recent study showed, more than two-thirds of patients survived that first year, most without the need for a new heart.

She was started the same day on high doses of prednisone and cyclosporine, an immune-suppressing medicine often used in transplant patients. Since her immune system was almost completely incapacitated by the treatment, she also took antibiotics to prevent infections.

Over the next six months, she did well. She still couldn't quite manage the brisk three-mile walks she used to take with her friends, but she was up to the two-mile mark and feeling confident she'd make it to three by the end of the year. She will never regain the heart muscle destroyed by the disease, but improving what's left will help her make up for what's gone. She continued to take immune-suppressing medications, but Stewart was tapering the dosage.

She and her husband got new wedding rings. Her other losses won't be as easily replaced, but she was determined to get back to where she was—a state of health she never fully appreciated before.

Pulseless

Don't worry," the teenage girl said as she lowered herself to the ground. It was her last day at her suburban New Jersey high school, and she and her friends were in the lounge, studying for their very last exam.

"This happens to me," she mumbled as she lay back and closed her eyes. "Don't call an ambulance." A friend grabbed the girl's wrist, searching for a pulse. Blood began to bubble through the girl's lips and drip from her nose. The friend rolled the girl onto her side, and an enormous amount of dark red blood gushed out onto the carpet. "Call 911," she shouted to the girl's other stunned friends nearby. "And get the nurse."

Suddenly the friend couldn't feel a pulse. She felt all around the neck, searching for a carotid artery. Still nothing. She rolled the girl onto her back and began cardiopulmonary resuscitation. She was taking a class to become an emergency medical technician, and she pressed her hands, one over the other, deeply and rhythmically right over the girl's sternum to the mental beat of the song "Stayin' Alive"—just as she'd been taught. The nurse arrived with a defibrillator. She applied the sticky pads to the young woman's chest, then attached them to the machine. A recorded voice instructed the nurse to give the girl a shock. The girl's body stiffened as the three thousand volts shot through her. Continue CPR, the machine instructed. The girl received a total of three shocks, alternating with CPR, before an ambulance arrived to take her to a nearby hospital.

The girl had been in the hospital a few months earlier. Around Christmas, her mother found her on the floor, her head encircled by a halo of blood that had clearly come from her mouth and nose. Her terrified mother immediately called 911. "Is she breathing?" a dispatcher asked. She placed her hand on her daughter's rib cage. "I think so," she said.

An electrocardiogram done at the local emergency room, that first time, was abnormal. Blood proteins called troponins were elevated, a sign that the heart muscle had been damaged. Clearly something happened to this young patient's heart. Yet the cardiac monitor showed a regular rhythm. And an echocardiogram—a picture taken with sound waves—showed that the muscle was working normally. Over the next few days, all signs of heart damage disappeared.

Back at home, the girl wore a heart monitor for a month, which would document abnormalities that could provide a clue as to what happened. The data were completely normal. By spring, both mother and patient began to feel a little optimistic. Maybe whatever was wrong had passed, and she was okay. Then came the episode at the high school.

The ambulance took the girl to Newark Beth Israel Medical Center. She was conscious by then, and the cardiologist explained that this episode was probably caused by the same problem that caused the episode before Christmas—her heart had started to beat abnormally and then just stopped. He wasn't sure why and, worried that the same thing could happen a third time, recommended that they surgically implant a defibrillator in her chest. Should her heart stop again, the device could save her life. Everyone agreed, and the tiny device was inserted.

The rest of the team, meanwhile, had to figure out where all the blood was coming from. Usually, profuse bleeding from the mouth comes from the upper gastrointestinal tract. But when a scope was passed through her mouth and esophagus into her stomach, it did not show anything that would have caused significant bleeding.

The other possible source was her lungs, and to look there they would need a CT scan of the girl's chest. She had been in the hospital for nearly a week at this point. She'd had many tests, many needle sticks, and no answers; she was worried and frustrated. Another test—during which she would be uncomfortable and cold and would have one more needle stick (for the contrast dye)—seemed more than she could bear. Even as she was transported to the scanner, she wept, asking her mother if the test could be canceled. "I can't do it," she wailed. But she did, breathing when instructed and holding her breath as much as she could between sobs.

Dr. Tej Phatak was both a pediatrician and the radiologist helping to manage her care. Through the control room window, he saw the girl's tear-streaked face disappear into the scanner, then watched the pictures click into place on his monitor. Suddenly he saw an irregularly shaped splash of brightness in the bottom of her right lung, in a place that should have been mostly dark. It looked like a tangle of vessels. He recognized it immediately. It was a pulmonary arteriovenous malformation, or PAVM, an abnormal connection between a pulmonary vein and an artery. Veins are thin-walled structures that carry slow-moving blood and expand or contract depending on how much blood is flowing at that moment. Arteries are thick, muscular vessels built to hold up under the fast-moving, high-pressure blood that comes out of the heart with each beat. The structure he saw was too big to be a vein and too misshapen to be an artery. It had to be a PAVM—and they are prone to tear, which can result in dangerous bleeding.

Phatak rapped sharply on the window to get the attention of the patient and her mother. "I know what you have!" he shouted. He hurried into the scanner room to explain what he saw and what it meant. What she needed, he told them, was a procedure to close off this vessel so that it would never bleed again. The patient was sent to Yale New Haven Hospital, and the tangle of vessels in the lower part of her right lung was closed up with loops of fine wire, about the width of a human hair. Blood would clot around the coils of wire, essentially sealing off the vessel for good.

For most patients, a single diagnosis is all that is needed to answer the question "What's wrong with me?" Not so in this case. These abnormal pulmonary vessels indicate a diagnosis that, in turn, often signals the presence of yet another rarity—a disease known as hereditary hemorrhagic telangiectasia, or HHT. This disorder, usually passed from parent to child, causes abnormal blood vessels to form, most frequently in the stomach, lungs, nose, liver, or brain. In addition, most patients (though not this one) with HHT develop red freckle-size spots known as telangiectasias, tiny malformed vessels located throughout the body. These abnormal vessels can cause serious bleeding, as in this case, as well as infections or even strokes. Patients with HHT tend to have frequent, even daily, nosebleeds—as this girl did—and can experience blood loss through the stomach or other parts of the gastrointestinal tract, in their lungs, or even in their brains.

A genetic test confirmed that she had one of the mutations that cause HHT. Because neither of her parents have the disorder, it's a mutation that occurred in this girl's genome just after conception. It was important to confirm the diagnosis, because if she had children, they would have a fifty-fifty chance of inheriting the mutation and the disease.

It's still unclear what made the girl's heart develop that abnormal, potentially deadly rhythm. It's not part of the usual manifestations of HHT. And her cardiologist wonders if she has a heart problem in addition to her HHT. She and her mother hypothesize that the PAVM and bleeding somehow triggered the arrhythmia. And it's true that she had no arrhythmias in the six years after her PAVM was treated.

The young woman is reminded of her disease by her near-daily nosebleeds—a common problem among those with HHT. But she hasn't let them slow her down. She graduated from college and has been hard at work in her career ever since.

PART VII

Strange Rash

Red Scare

"Whoa! That is definitely not poison oak," Dr. Walter Larsen blurted as he entered the exam room in his Portland, Oregon, office. The patient smiled ruefully. "I told you," she said to the doctor. The fifty-six-year-old woman had seen Larsen that Tuesday, two days earlier. Then she was concerned; now she was scared. She looked down at her hands and arms. Her fair skin was nearly hidden by vicious-looking streaks of red. "And it's everywhere." She lowered the office gown to reveal scarlet lines crisscrossing up her arms, across her neck, down her back, chest, and abdomen. Larsen leaned in close to get a better look at the patient's red-streaked skin. He didn't know what this rash was, but it sure had become a lot uglier in just a couple of days.

The patient first saw the rash on Monday, and at that point, it was just a rash confined mostly to the back of her hands. It didn't really hurt or itch. By the end of the day, though, the rash had become redder and angrier-looking.

Overnight, tiny blisters formed over the red regions. When her sister saw her hands, she was concerned. "You've got to see a doctor about this," the sister urged. The patient was reluctant; she'd been laid off from her job at the local power company and now had no health insurance. Still, the rash on her hands looked pretty awful. And now it was painful as well. She called Larsen's office, and they fit her into his schedule later that day.

At the first visit, Larsen immediately suspected that it was some kind of allergic contact dermatitis, probably from a plant like poison oak. Although it was late in the season for that kind of rash, this was a condition he knew a lot about. He had helped write a book on contact dermatitis. He asked the patient if she had been outdoors within the past few days. She told him she visited a friend's farm and picked Swiss chard over the weekend, but she didn't see any poison oak. Nevertheless, that seemed to cinch it—at least for the doctor.

The patient wasn't convinced. She'd never had a reaction to poison oak before. And besides, wasn't a poison-oak rash famously itchy? This rash was tender to the touch but not itchy at all. Still, the setting was right—she was out among plants—and the rash looked enough like poison oak at that first visit that it was hard for Larsen to discount it. He gave her a steroid cream to use and suggested that she come back in a couple of days so he could see how she was doing.

Now, two days later, she was back, and Larsen was stumped. The little blisters he noted on her hands initially had hardened, and the red streaks were much darker, almost purple, and raised. The way those welts now streaked across her neck, back, legs, and abdomen made it look as if she had been flogged. Indeed, she told the doctor, she had started wearing gloves and long sleeves to hide the unsightly markings.

Had the rash started itching? The red streaks looked like excoriations from vigorous scratching, but she told him that she hadn't been. Besides, she had this rash on the middle of her back, where she couldn't even reach. The doctor pulled a capped pen from his breast pocket and drew it lightly down her back, leaving a faint red line. As he did so, he explained that some people have a condition known as dermographism, in which the skin has an allergic reaction to being touched. With these patients, applying pressure as he just had could cause a red welt, like the ones she had on her body. He waited. The mark faded.

Had she started taking any new medicines recently? An allergic drug reaction could cause this kind of whole-body rash—though he had to admit he'd never seen one like this before. She shook her head: no new medications. There was no fever, no symptoms other than the rash? None, she told him. That made an infection unlikely.

"Okay—it's time to call in reinforcements." Larsen asked if he could bring in a couple of colleagues and then disappeared from the room, returning a few minutes later with two of his younger partners. After a long moment, one of the partners, Dr. Michael Adler, broke the silence. He asked the patient whether she had eaten any shiitake mushrooms recently. The question surprised her. "How did you know?" she asked. On Friday, three days before the rash appeared, she was offered a sample of shiitakes cooked in oil and garlic at her local grocery store. They tasted fine, maybe a little chewier than usual, but she enjoyed them.

The young doctor thanked her, and then the three walked out of the room without telling her anything more. Finally Larsen returned. "We think this is a classic reaction to raw or undercooked shiitake mushrooms," Larsen told her.

Shiitake dermatitis, as it's known in medical jargon, was first described in 1977. Since then, it has been frequently reported in Asia, though rarely, if ever, in the United States. The rash is thought to be a toxic reaction to a starchlike component of the shiitake mushroom. This component, known as lentinan, breaks down with heat, and so this reaction is seen only when the mushrooms are eaten raw or partly cooked.

"So am I allergic to these mushrooms?" the patient asked. Well, it's not a true allergy, Larsen explained. When someone is exposed to a substance and has a bad response, it's considered allergic only when the immune system causes the reaction. Then you'll get hives or swelling or occasionally anaphylaxis. But when people who get this rash are tested, there's no sign of an immune response, so it's considered a toxic, not allergic, reaction. The current thinking is that something in the lentinan triggers blood vessels to dilate and leak small amounts of inflammatory compounds just beneath the skin.

Not everyone has this kind of violent reaction to raw shiitakes. In one study, just over five hundred patients were exposed to an intravenous version of lentinan. Nine developed this streaky rash. The other patients had no response. Perhaps that's lucky for them, because this same component is thought to have important health benefits. Studies suggest that lentinan may be helpful in preventing diseases ranging from cavities to colon cancer. Why it creates these whiplike streaks in some is not well understood. A rash with a similar pattern has been linked to bleomycin, a chemotherapeutic medication.

Larsen ordered a biopsy of the rash to make sure they weren't missing anything, and he instructed the patient to continue to use the steroid cream at home. The cream helped, but it took weeks for the rash to fade completely.

Larsen recommended she avoid uncooked shiitakes. "I'm never going to touch another shiitake," the patient told him. "I don't care if they are good for you. One of these rashes was enough."

So how was Adler able to recognize this obscure rash when his older, more experienced colleague was not? I put the question to Adler. He laughed. "I'm not usually the guy who gets the off-the-wall diagnosis," he told me. "It was really just luck." He had read a case report of a patient who developed this crazy-looking rash after eating shiitakes. The picture in the journal was so striking that as soon as he saw the patient, it all came back in a flash. "That's what is so great about working in a group. When you get stumped, you just call for help, and chances are, one of these guys will know the answer. It's like doing the crossword puzzle with a friend. With any luck, the other guy fills in your gaps in knowledge. This time I got to be that guy."

Near Death at the Hands of Doctors

The loud crash seemed to shake the house. The woman hurried toward the source of the noise and found her fifty-seven-year-old husband on the floor of his TV room. Her twenty-three-year-old son, who was studying to be a nurse, had pulled his father out of the bathroom where he'd fallen. Her husband's face was swollen, the skin darkened to an unnatural reddish purple. His eyes were open but unseeing. A strange gurgling noise emanated from his open mouth. The remnants of the pitcher of laxative he was taking in preparation for the next day's colonoscopy sat on his desk.

"Get the EpiPen," her son shouted. He knew they had one from the only other time his father had had this sort of reaction. Call 911, he instructed her once he pressed the automatically triggered syringe into his father's thigh.

When the medics arrived, the man's blood pressure was

so low it was hard to measure. His breathing was ragged and loud—his trachea dangerously narrowed by swollen tissues. The EMTs were debating whether to cut a hole in his trachea, to allow the air to get to his lungs, when the man opened his eyes. His breathing quieted as the epinephrine finally did its work. He was taken to the Williamson Medical Center in Franklin, Tennessee, a suburb of Nashville.

After more epinephrine and IV fluid, the man started to feel better. He was discharged hours later with instructions to contact the allergy clinic at Vanderbilt. He was allergic to something that nearly killed him; he had to find out what it was. The next time he might not be so lucky.

Although he had some trouble with penicillin when he was a kid, as an adult, he had never had any allergies—until two years before this incident. Back then a doctor had suggested a steroid shot for some arthritis in his neck that had not responded to Tylenol or physical therapy. An injection of a steroid medication into the spinal area, the doctor said, could reduce the inflammation and help with the pain.

He went to a specialist for the shot. It was just a pinch—not painful at all. But as he stood up to put his shirt back on, he suddenly felt hot all over. "I don't feel right," he announced, alarmed. His hair was drenched in sweat. He felt pins and needles all over his body. Looking down, he saw that his arms and hands were covered by angry red welts. He was dizzy, and the world suddenly looked like a TV screen with bad reception. He felt a quick sting in his thigh. Epinephrine, they told him.

The next thing he knew, he was in an ambulance. Then, suddenly, he was in the ER, with an IV in his arm and his wife at his side. Someone gave him another shot of epinephrine and more IV fluids. He felt worse—his body gripped by spasms that nearly flung him off the bed—and then, eventually, better. After several hours, they said he could go home.

He had an allergy to the steroid, he was told. It didn't make sense, because the body is naturally awash with steroid hormones. But his reaction was clearly allergic, and of the most severe type: anaphylaxis.

The anaphylaxis scared him. But when it didn't happen again, he relaxed a bit. Still, when his doctor's office called two years later to schedule a colonoscopy, he was wary of ingesting anything new. He asked about the laxative he was to take in preparation for the test. He was reassured that although this medicine had a different name, it was basically the same one he took seven years before, for his first, uneventful colonoscopy. He took a couple of swallows, then waited. Within minutes, his mouth started to itch, and the strange pins-and-needles feeling that preceded the welts started. He took some Benadryl, and slowly the symptoms subsided. He reported the problem to his GI doctor. He still wanted the colonoscopy but asked for a different laxative to prepare for the exam.

He looked at the new package closely. It had a different name and was made by a different manufacturer. Relieved, he drank down the first big glass. But within minutes, he started to experience the same awful symptoms he had after the steroid shot. He took two Benadryl, but it wasn't enough. He was drenched in sweat, and he had a ringing in his ears so loud he could barely hear. He could feel the sting of the welts rising all over his body. The skin on his face was ablaze and tight. Suddenly he felt as if a trapdoor opened under his feet, and he disappeared into darkness.

When he came to, he heard the EMTs discussing whether they would need to cut a hole in his airway. The noisy rattle of his own breath sounded thunderous, and he was afraid. But his breathing improved, and he was once again hustled to the ER by ambulance.

After this third reaction and second trip to the ER, he took the doctors' advice and called the allergy clinic at Vanderbilt. The earliest appointment he could get was weeks away. Impatient and worried, the man started his own investigation. From the first doctor, he got the name of the steroid medicine injected into his spine. It was something called Depo-Medrol. He looked up the colonoscopy prep medication that had given him the same reaction, GaviLyte-C. And the one that had made his mouth itch, MoviPrep.

When he compared the three products, the only ingredients they had in common were salt (sodium chloride) and something called polyethylene glycol. PEG, as it's called for short, is an inert chemical used in both industry and medicine, as a lubricant and filler in products ranging from hand lotions to hair spray to gel caps to pills. And, as this patient discovered, PEG is also used in some steroid preparations and some laxatives.

Armed with this information, he followed up with the doctors at Vanderbilt. Dr. Cosby Stone, a physician who specializes in allergic reactions to medications, introduced himself to the couple and invited the patient to tell his story. "I'm not trying to tell you how to do your job or anything," the man started, "but I'm pretty sure I'm allergic to PEG, polyethylene glycol."

Stone was amazed. Few patients come in linking an allergy to such an obscure product. The patient described his experiences with the three drugs. There were only two ingredients all the drugs had in common, PEG and salt, and he ate a lot of salt, so it couldn't be that.

Stone asked his adviser, Dr. Elizabeth Phillips, to join them in the exam room. She heard the man's story, and then the two doctors left to discuss the case. Could PEG cause this kind of severe allergic reaction? A review of the literature turned up a handful of similar cases. Still, it would be important to make sure that it was PEG causing this life threatening reaction. They returned to the room and congratulated the man on his sleuthing abilities.

Over the next few weeks, they did the testing. He had a severe allergy to PEG as well as one of its chemical cousins, polysorbate 80. The man worked for the regional power company and was often exposed to an industrial version of PEG. In genetically predisposed individuals, this kind of repeated exposure can lead to an allergic reaction.

Before he used any new product or drug, the doctors instructed, he would need to do a careful inspection of the ingredients. And he should get a medical alert bracelet to make sure others knew of his allergy so he didn't receive any products containing PEG or polysorbate 80 by accident.

Recently the patient noticed that a new hand lotion bought by his wife caused his hands to itch. He looked at the label. Sure enough, one ingredient, way down the list, was PEG. Stone had warned him. The chemical is everywhere.

For Stone, this case represented so much of what he loves about his job—the chance to sit down with his patients and really hear their stories. And to follow the advice of the celebrated twentieth-century physician Sir William Osler—to listen to the patient and let him tell you what he has.

Old-Fashioned Skin

"I feel a swimming in my head," mumbled the voice on the phone. Dr. Stephanie Pouch, a resident in her second year of training, wasn't sure what to make of this gentleman's strange complaint. Was he feeling dizzy? He wasn't sure; all he could say was that his head was "swimming" and that he had almost passed out. Uncertain of the cause or even the nature of the complaint, Pouch sent the man to the emergency room. She was on call that day at the University of Chicago Medical Center. She would figure it out when he got there.

Once Pouch found the sixty-eight-year-old man in the busy emergency department, a quick glance at his chart explained his vague complaint: His blood pressure was so low it could barely be measured. Whenever he tried to stand, his blood pressure dropped even further, and his head began to swim. A look at the patient himself explained the dangerously low blood pressure: He was severely dehydrated. His eyes were dull and apathetic; his dark skin hung off the bones of his face as if it were a size too large. Beneath a graying, well-trimmed mustache, his lips were dry and cracked, and he passed an equally dry tongue across them frequently.

He was started on intravenous fluids. That would certainly help. But what happened to bring him to this state? The patient was a man of few words. But slowly, with the help of the patient's wife, Pouch was able to put together his story. For the past several months the patient was plagued by severe diarrhea. He was in the bathroom five to ten times a day. And the pattern repeated itself at night. He couldn't remember the last time he had an uninterrupted night's sleep. He had no pain, no fever or chills, just these endless trips to the bathroom.

As the patient spoke, Pouch's eyes were drawn to his hands. They were covered by strange stripes of dark, thick, rough skin that started at the knuckles and extended all the way down the fingers. She gently turned one of the patient's hands to look at the palm and found more of the same. He'd had the rash a long time, he told her, for weeks, maybe months. Pouch found the same rash on his back, chest, and feet.

The rash was like nothing Pouch had seen before. Still, it was the diarrhea—not the rash—that brought the patient to the hospital. She forced herself to focus on the problem: What could be causing this persistent and profuse diarrhea? Pouch paged through the patient's thick chart. He had a lot of medical problems: diabetes, atherosclerotic disease (also known as hardening of the arteries), and, usually, high blood pressure. This combination of diseases could cause diarrhea by limiting the amount of blood that reaches the intestines and starving the tissue. There were many other possible causes of severe diarrhea: Infection was one; cancer another. Certain tumors can cause diarrhea by producing too much digestive hormone.

Pouch sent off samples of the patient's stool to look for evidence of infection and blood for overproduction of hormones. The patient also needed an ultrasound of the blood vessels that feed the gut to see if blood flow was compromised.

The next morning on rounds, Pouch presented the patient to the attending physician, Dr. Vineet Arora. The more experienced Arora was worried about the patient's prolific diarrhea, but she was also struck by the unusual rash. Could they be part of the same disease process? Patients like this man are the most difficult to assess, Arora later told me, because they have many medical problems, and it's difficult to distinguish the foreground (the disease) from the background (the complicated and sometimes abnormal baseline state that now represents the patient's normal condition).

There were some important diseases to consider that could cause diarrhea and a rash. Celiac disease—a sensitivity to a component of wheat known as gluten—can cause both. And the rash in celiac disease, unlike most rashes, can spread to the palms of the hands and soles of the feet. Zinc deficiency could cause both. So could a number of B-vitamin deficiencies. Pouch quickly ordered a series of blood tests for these deficiencies.

Over the next several days, the patient improved significantly. His diarrhea slowed, his blood pressure rose, and he could sit and stand without the dizziness that took him to the hospital. Meanwhile, test results dribbled in but provided no real answers. The ultrasound confirmed the hardening of the arteries but showed adequate blood flow to his intestines.

There was no evidence of infection. He didn't have celiac disease. It wasn't zinc deficiency. After nearly a week, the patient was better and was sent home, though the team still didn't know what made him sick.

A couple of days later, Arora and her team got their answer—or at least part of it. The blood tests they ordered revealed that the patient had a severe deficiency of vitamin B_6. Initially, Arora was flummoxed. Vitamin B_6 deficiency is rare in this country, and while it can cause pain in the hands and feet, it causes neither a rash nor diarrhea. Finally, she hit pay dirt: deficiencies of this essential nutrient caused the patient to develop a condition known as pellagra. First described by eighteenth-century European physicians, the name pellagra comes from an Italian description of its most common symptom: the "rough skin" that Arora and Pouch were first struck by when examining this patient.

For centuries, the disease was thought to be caused by an infection, but we now know that pellagra comes from a deficiency of niacin. If an individual is not ingesting niacin, the body can create it, but it needs vitamin B_6 to do so. In medical school we are taught that pellagra is characterized by the four Ds: diarrhea, dermatitis (rash), dementia, and death. This patient had two of the four.

If this vitamin B_6 deficiency explained the rash and diarrhea, what explained the vitamin B_6 deficiency? Again, Arora wasn't sure. Further reading led to the answer. The patient was taking hydralazine, a blood-pressure medication that had the side effect of eliminating vitamin B_6 from the body. Hydralazine was an old blood-pressure medication that slipped from use as newer, easier-to-take medications were developed. But a 2004 study suggested that hydralazine might be particularly useful in African Americans. This new information brought the old medication to the current generation of doctors—and African American patients like this one. That hydralazine can also cause a vitamin deficiency was common knowledge in this medication's first life but seems to have been forgotten in its reprise.

Now the story was beginning to make sense: The hydralazine caused the vitamin B_6 deficiency, which, in turn, led to the niacin deficiency and pellagra. The patient's diarrhea caused the low blood pressure. While he was in the hospital, the doctors were not giving him hydralazine because his blood pressure was low. Without the hydralazine the patient was able to absorb vitamin B_6 and make niacin. By the time he was discharged, the diarrhea had improved strikingly. Arora contacted Dr. Kevin Thomas, the patient's primary care physician, who immediately started him on vitamin B_6 supplements. The diarrhea resolved completely within a week; the remarkable rash disappeared over the next two weeks.

Arora and Pouch presented this case to other doctors and found that few of them were aware of this side effect of hydralazine. "Why don't we know this anymore?" Arora asked me, amazed. "If this medicine is going to be used again, then doctors definitely need to know about this problem."

Red and Sore All Over

The fifty-five-year-old woman barely recognized the face that stared back at her in the bathroom mirror of the emergency room. Her eyes were so swollen that she looked at her distorted reflection through tiny slits. Her skin was puffy and red. And her chest was dotted with several scarlet-colored blotches.

She was camping in the mountains of Vermont with her husband, her cousin, and her cousin's husband. They had arrived the previous night in an RV at a campground in the Green Mountains. It was raining, but they had a pleasant dinner inside the camper, then went to bed. When the woman awoke the next day, she knew immediately that she had a fever. Her eyes felt irritated, her skin itched, and when she brushed her teeth, the toothpaste she spat into the sink was red with blood. Even urinating was painful.

Her cousin's husband, a physician, looked into her swollen eyes and saw that they were red and irritated. Then he looked into her mouth. On her tongue and along the inside of her cheeks were blisters filled with a dark—almost black—liquid. Blood, he realized.

"You're going to the hospital," he announced. He didn't know what this was—he was a radiologist and hadn't seen anything like this before. But he was certain that she needed to see someone who had.

The drive to the hospital was the worst two hours of the woman's life. The late-afternoon light hurt her eyes. Her skin felt tender and itchy. Her head ached, and the motion of the car made everything worse. She almost cried with relief when they finally reached Brattleboro Memorial Hospital.

Doctors, nurses, and technicians zipped in and out, peppering her with questions, hooking up IVs, taking blood. By evening she was able to sit up. Antihistamines had calmed the itchy rash. A middle-aged woman with a kind but no-nonsense manner introduced herself as Dr. Teresa Fitzharris. At her prompting, the patient began telling her story—how she awoke feeling so sick after going to bed feeling fine.

Actually, she told the doctor, except for the day before, she hadn't felt well for nearly two weeks, ever since she and her husband camped out in western Vermont. On that trip, she got a bunch of black-fly bites, and each bite swelled into an enormous welt. That had never happened before.

The following week she traveled to see her niece on Long Island. When she got home from that trip, she had an awful headache. She took one ibuprofen and went to bed. The next day the headache was gone, but she had a fever of 101. She kept taking ibuprofen for the fever but thought it was strange that, except for that, she didn't really feel sick. The fever lasted a couple of days, and she was worried that she wouldn't be able to go on this trip with her family. But the day they were to leave, she felt good—no fever, no headache. Now the fever was back, and she had never felt sicker.

The patient had a fever of 101.6. Her heart was beating rapidly. Her eyes were swollen and deeply bloodshot. Her lashes were crusty with a yellow discharge. Her throat was red, and there were multiple blood-filled blisters on her tongue and cheeks. Her neck, chest, abdomen, and back were now covered with a bright red, bumpy rash that was very itchy and a little sore.

Fitzharris reviewed the lab results. They were all normal. The chest X-ray was clear. Cultures of the blood, urine, and discharge from the eye were pending. She called the infectious disease doctor, David Albright, who was on call that night, at home, and she went over the case with him.

It was clear this was some kind of febrile illness, but which kind? It might be a virus. Coxsackievirus often attacks the skin and mucus membranes. Adenovirus could cause a fever and infection of the eyes and throat. Or could she have started out with a virus, then had an additional bacterial infection? Given her recent outdoor exposure, this could also be some kind of insect-borne illness like Lyme disease or Rocky Mountain spotted fever. The patient hadn't seen any ticks, but then, many patients don't.

Albright wondered if it was an infection at all. The runny eyes and the itchy rash could be an allergic reaction. The patient had no history of allergies, but they can develop at any age. Fitzharris started the patient on antibiotics to treat the most likely infections.

When Albright saw the patient the next day, the rash was bright red. The welts on her back had merged into a single huge, bumpy red blotch. Her fever had come down, but she remained achy, itchy, and uncomfortable. He wasn't sure what this was, but he was certain he had never seen it before. He immediately called Dr. Jorge Crespo, a dermatologist. He recounted the course of the patient's illness, the exam, and the diagnoses that he and Fitzharris considered.

The bloody blisters in the mouth made Rocky Mountain spotted fever unlikely, Crespo told him. And the dense red rash made the common tick-borne illnesses Lyme and anaplasmosis unlikely. Crespo said he would have to see it to be sure, but from the description, he didn't think this was an infection at all. He would put his money on a severe allergic reaction known as Stevens-Johnson syndrome. In this potentially life-threatening disease, something—sometimes an infection but more commonly a medication—will trigger the body to attack the deepest layers of the skin and mucus membranes, causing blistering and sloughing of the top layers, almost like a severe burn. He would be in to see her soon.

Ibuprofen and other nonsteroidal anti-inflammatory drugs are among the most commonly used medications and are therefore a common cause of side effects. Up to 7 percent of hospital admissions are related to adverse effects of drugs generally and, of these, nonsteroidal anti-inflammatories are responsible for more than 10 percent—usually affecting the gastrointestinal tract or kidneys. When it comes to Stevens-Johnson syndrome, they are the third-most-common drug to trigger it. (Bactrim and other sulfa-based antibiotics are more common, and some antiseizure and gout medications have also been linked to Stevens-Johnson.)

One look at the patient's skin, eyes, and mouth convinced Crespo that this was indeed Stevens-Johnson syndrome. After talking with the patient, Crespo thought she was probably reacting to the ibuprofen she took. She took it first for a headache. When she developed a fever, she did not recognize that as a symptom of this rare but devastating allergic response, and she (logically) continued to take it for several days to reduce her fever.

Although there is debate about how best to treat Stevens-Johnson syndrome, Crespo recommended the use of steroids to damp down the immune system's misdirected attack on the patient's skin. An ophthalmologist was consulted because Stevens-Johnson can cause blindness. The patient remained in Brattleboro Hospital for a couple of weeks, but she didn't fully recover for months. Even now, ten years later, the patient cannot forget her ordeal. The inflammation of her eyes was so severe that it scarred her tear ducts. Ever since her hospitalization, she has had to put saline solution into her eyes every few minutes, because she is no longer able to make tears. Those drops are second nature now—but remain a daily reminder of her experience.

I called Crespo to ask him how he was able to make this diagnosis so quickly. Was it the history? Was it the exam? "Dermatology is all about seeing," he said. "It's all visual." But in this case, the diagnosis was made over the phone. "It was still visual," Crespo said. He was simply using someone else's eyes.

A Black Thumb

The seventy-two-year-old woman carefully loosened the bandage she had wrapped around her thumb. As she slowly revealed the injured finger, her daughter gasped. The top third of what should have been the fleshy part looked eaten away. The flesh that remained was black and hard. A foul odor emanated from the wound. "Yesterday I thought there was a dead mouse in my living room," the woman told her daughter. "Then I realized it was my thumb." She couldn't do anything with that hand now, not even work in her garden. It was very upsetting.

The daughter wondered: Was this gangrene? Was her mother going to lose her thumb? She took out her phone and snapped a picture of the digit. She and her mother's internist worked in the same medical center in nearby Joplin, Missouri, and he needed to know how serious this was.

The doctor had seen the rash before. The patient had come to his office a few months back when her hands first swelled up. The next time he saw her, just a few weeks ago, the skin on her hands had red, painful blotches. But this was new, he told the daughter after seeing the pictures. He didn't know what her mother had. After a moment of silence, he made his recommendation: She should take her mother to the Mayo Clinic in Rochester, Minnesota, nearly six hundred miles away. "Don't make an appointment," he told her. Take her to the emergency room. They'd figure it out. He was sure of that.

Early the next morning, mother, two friends, and her daughter set off on the ten-hour, four-state trek north to the Mayo Clinic. By the time they arrived, the mother could barely walk to the door. In the waiting room, she suddenly began to shiver. She was so cold, she told her daughter. Her teeth chattered. She had a fever. To her daughter, she seemed disoriented and confused. Clearly, they'd brought her in just in time.

It was past midnight when the patient was admitted and settled into a hospital room. Dr. Daniel Partain, the doctor in training assigned to her care, introduced himself. She'd had psoriasis for many years. It was scaly and sometimes pretty itchy, but she was able to treat it with a steroid cream. Then she developed this awful joint pain. It was mostly in her hands, wrists, and elbows, but other joints occasionally joined in the cacophony of redness, swelling, and pain. A few years earlier she started to see a rheumatologist, who told her she had something called psoriatic arthritis—an aggressive type of arthritis caused by the body's immune system, which mistakenly attacks itself. Left untreated, this disease can destroy bone. He started her on strong immune-suppressing medicines, and she did feel better.

She'd been on those medicines for four years and was

doing fine. Then, three months earlier, she started having episodes of swelling, mostly in her hands and arms. And then these ugly red patches appeared. Her joints didn't hurt, but the rash sure did. The rheumatologist attributed it to a flare-up of her psoriatic arthritis and increased her immune-suppressing meds. When that didn't help, he switched her to more powerful drugs. But her hands, especially her thumb, just got worse.

The woman told her story cheerfully enough, but it was clear to Partain that she was in pain. Her right hand lay mostly immobile on her lap, swollen and covered with red blotches. Her thumb was shockingly ugly, the raw flesh on the tip covered by a thick black scab. There were other areas of scaly, deep red skin on both her arms, but nowhere else.

The young doctor wasn't sure what was going on. If the psoriasis component of her psoriatic arthritis was causing this rash and black thumb, then why weren't her medications helping? Maybe she had an infection as well; because the meds she took suppressed her immune system, the body's most important defense against invading organisms, she was vulnerable to all kinds of bugs. Partain ordered antibiotics just in case. He asked that a rheumatologist and a dermatologist come by later that day. He gave the woman something for the pain, and then he went to see his next patient.

In the morning, Dr. Ruth Bates, the senior resident on the team, listened with interest as her intern told her about this new patient. She praised him for his thoughtful approach. When they entered the room, the patient unwrapped the bandage to give the team a chance to see the lesion that brought them together. They examined the patient and then left, promising to come back after lunch.

As soon as they were in the hall, Bates turned to her intern. "I think I know what this is," she told him. "I've seen it before." Two years earlier, when she was an intern, a man came in with a sore just like this one. Like this woman, he had been on immune-suppressing medications. They tested him for everything, and it turned out that he had something uncommon in Minnesota—he had been infected by a fungus called *Histoplasma capsulatum*, resulting in a condition called histoplasmosis. Bates thought it was likely that this woman had it as well.

There are approximately 1.5 million different species of fungus on Earth, but only about 300 make people sick. Many live in soil and cause infection when inhaled. Different fungi are common in different areas of the country. Histoplasmosis is one of the most common endemic fungal infections nationwide, seen primarily in the southern and central United States—around the Mississippi and Ohio River Valleys. The patient's home was in a region where the fungus could be found. The organism is carried by infected bats or birds and deposited into the soil in their waste. When Bates went back to the patient's room, she asked whether she'd had any exposure to soil or birds. Yes, she was an avid gardener—and she had many bird feeders.

The doctors ordered tests of blood and urine to look for the fungus. The dermatologist took biopsies of the red plaques on her hands. The tests confirmed infection the following day, and she was started on an intravenous antifungal treatment.

Most people who get histoplasmosis never know it. The infection is either asymptomatic or causes symptoms mild enough that patients don't seek medical attention. Among those who have symptoms, most have a disease that is limited to the lungs, like bronchitis or pneumonia, that usually doesn't need to be treated.

Not so for this patient. CT scans confirmed the presence of infection throughout her body, including in her chest and abdomen. The immune-suppressing medications she was taking for her arthritis had created a welcoming environment for the fungus. As she took more and stronger medications, her immune system was further weakened, and the fungal infection only got worse.

Recovery for this patient was slow and difficult. Every day, she was given a bright yellow intravenous antifungal medication, which she called her Mountain Dew from Hell. Each infusion caused her heart to race and her blood pressure to spike. She had constant diarrhea. Despite treatment, the fungus ate through her gut, and she had to have a foot and a half of her small intestine removed in an emergency operation. In all, she spent four long weeks in the hospital and a few more in rehab back in Joplin. Even after all that, she had to take an antifungal medication every day for a year.

Nearly three years out from that long trip to Mayo, she had made adjustments. She will never be able to take any immune-suppressing medications again. Her arthritis bothers her a lot more, but she says that taking extra-strength acetaminophen helps. Even though she's off immune-suppressing medications, she doesn't take any chances: Her bird feeders are gone. And every spring and summer, she has her six grandchildren do all her planting for her.

Line Dancing

"Wow! What's that on your arm?" a young dancer exclaimed in alarm. The forty-one-year-old dancer and choreographer looked at the part of her arm the other dancer was pointing to. A bright red line snaked down the side of her forearm, from wrist to elbow. She'd been completely unaware of it; it didn't hurt or itch, so she dismissed it. For the past several days, they'd been rehearsing in an old barn on Martha's Vineyard converted into a performance space. Maybe she'd scraped it on something. In any case, she didn't have time to worry about it, she told her friend. They had only four days until they were supposed to perform this piece.

The dancers had come to this beachside idyll for a month-long residency at an artists' retreat. Their days were divided between their creative work in the barn and the respite of the island's sandy shore. Because the funny red line didn't hurt much—just a little irritated—and because she felt well overall, the choreographer probably would have forgotten about it completely, except that people kept asking her about it.

Everyone seemed to have a theory. Some were certain it was poison ivy. Maybe, she replied, but it didn't itch. Others thought it looked like a jellyfish sting. Or an infection. Possible, she answered, but she expected those would hurt more. On the day of the performance, a staff member of the facility offered to take her to the urgent care clinic the following morning. She really should get it checked out, the woman urged. The choreographer was torn. On one hand, the mark on her arm was strange and mysterious. On the other, it wasn't bothering her, and the next day would be her last free day on the island and her final opportunity to enjoy the beach and the waves.

Still, she was curious. Surely one of the local doctors would be able to tell her what it was. And so, late the next morning she had someone drive her down the island to the urgent care clinic. The doctors there were busy when she arrived, but the nurse said she could look at her rash and at least tell the patient whether it was worth being concerned about. She followed the nurse into the exam room and pulled back her sleeve to reveal the red line, which was now raised and angry-looking. She heard the nurse inhale sharply. She really should to go to the ER, the nurse told her. She wasn't sure what it was, but it worried her.

In the emergency room, she was seen by a nurse and then a physician assistant. Both of them were puzzled. The nurse told the dancer she'd lived on Martha's Vineyard for over a decade and had never seen anything like it. The physician assistant wasn't sure what it was, either. The dancer left the

ER with a prescription for a steroid cream and was warned to come back if it got any worse.

Back at the performance barn, another staff member approached her. He'd sent a picture of her arm to his mother, who was an infectious disease specialist in Washington State. She'd shared the photo with her colleagues in her office, and while none of them were certain of what it was, the consensus was that it might be an atypical presentation of the erythema migrans—the rash seen in Lyme disease. She was, after all, on Martha's Vineyard, where Lyme disease was practically epidemic. The doctor suggested she take two weeks of doxycycline.

That seemed reasonable to the dancer. She'd certainly heard many warnings about checking herself for ticks, because the risk of Lyme disease was so high on the island. But it was Sunday, and the local pharmacy was closed, and they were heading back to New York City the next day. On Monday, the dancer sent a note to her primary care doctor, via the patient portal, asking for a prescription of the recommended antibiotic. She also sent a picture of her rash. Someone from the office called in the prescription, and she started taking it.

But the next morning, she got a call from her doctor. The physician was worried about the red streak. Maybe it was Lyme, but it didn't look like any of the Lyme rashes she'd seen. The dancer needed to see a dermatologist or an infectious disease doctor to really get a diagnosis.

By now, the dancer was getting a little worried. She called the office of Dr. David Bekhor, the infectious disease specialist her doctor recommended. He would see her that week.

Bekhor was a tall and tidy man, about her age. She briefly outlined her story. His first thought, as he listened, was that this was going to be sporotrichosis—a fungal infection often known as rose gardener's disease, because it lives on plants and is transmitted through breaks in the skin, like those caused by thorns. It can travel from the inoculation site—often on the hand—through the lymphatic system to form a lumpy red line that travels up the arm.

But when she showed him her arm, he dismissed that idea immediately. The vivid red line, about one to two centimeters wide, started right at the wrist on the pinkie side of the forearm and wended its way up the arm to the elbow. The affected skin was slightly raised and a little scaly. Bekhor recognized the rash immediately. "Were you drinking Coronas on the beach?" he asked. The dancer was put off by the question. What a thing to ask! No, she had not been drinking any kind of beer on the beach.

The doctor tried again: Had she been squeezing lime into any drink while on the beach? Sort of, she replied. She'd put lime juice in the water she took to the beach. Because she was dancing for so many hours each day, she was drinking lots and lots of water. And she would often squeeze some lime into the water for flavor.

That was it. She didn't have Lyme disease, the doctor said. She had what he called lime disease. The medical term is phytophotodermatitis. Lime juice contains chemicals known as furocoumarins, which cause a skin reaction when they're exposed to sunlight. At some point, while adding lime to her drink, some of the juice must have dripped down her arm. She'd washed her hands but not her arms, and the sunlight reacted with the dried juice to cause this skin reaction. The rash doesn't appear immediately. It takes up to twenty-four hours for it to emerge, and that makes seeing the link between the cause and the effect—the juice and the mark—more difficult. This isn't an allergic reaction, he added. It can and will happen to just about anyone who has the necessary ingredients: lime juice and sunlight.

Bekhor says he sees this rash a couple of times a year, usually in the summer, but also in the winter, during cruise season. He always asks if the patient has been drinking Coronas on the beach, because that so often is how beachgoers come into contact with lime juice. Most of the time, he told me, people are amazed—How did you know? they ask. He chuckled. The responsible chemical is found in other foods as well—grapefruit, lemons, celery, carrots, and parsley—but often the culprit is lime juice. Indeed, this reaction is sometimes known as margarita dermatitis. And the rash can be much worse than this patient's. People can get blisters and swelling, and they can be quite painful. She was lucky. The doctor explained that there was nothing to be done now. It would take a while—often months—for the rash to go away.

A couple of months after the Martha's Vineyard dance residency and the lime beach rash, the red line was still there. It had smoothed out, and the color was faded but visible. At least now, when asked about the streak up her arm, the dancer had a story to tell.

PART VIII

So Weak

A Terrifying Silence

The woman sat watching her three-month-old daughter lying limply in the hospital crib; it had been a terrible couple of days. Suddenly, the baby's plump cheeks lost all their color, and the quiet gurgle coming from the back of her throat fell silent. An alarm sounded, and a nurse hurried into the room.

The nurse glanced at the oxygen monitor, then quickly placed her tiny stethoscope on the infant's chest. She was barely breathing. The nurse grabbed the clear plastic suction tube kept at the bedside and eased it through the baby's lips, deep into her throat. Clear fluid bubbled back up the tubing, making a slurping sound, like the last drops of a drink going through a straw—saliva had pooled in her airway.

Finally the baby cried out feebly, like the plaintive mew of a kitten. The numbers on the oxygen monitor reversed their downward course as color crept back into her face.

The mother turned expectantly to the pediatrician who

had joined the nurse. She saw what looked like a flash of fear cross the doctor's face. Maybe they weren't going to be able to figure out what was wrong with her baby.

Just a few days earlier, the rosy, chubby three-month-old simply stopped eating. This was the mother's second child, and she knew that a baby's appetite could wax and wane, so she wasn't too concerned. When the child ate only once the next day, she started to worry. When the baby turned away from the breast again the following morning, and also refused a bottle, the mother called her pediatrician.

The doctor examined the baby, then sent mother and daughter to a specialist, suspecting an object might have become stuck in the baby's throat. When nothing was found, the pediatrician sent them to the local hospital for IV fluids and further evaluation.

At the hospital, the parents were told that the child probably had some kind of virus. She had no fever, and her vital signs and physical exam were normal except that she seemed a little weak. Blood and urine were sent to the lab for testing. The results were normal, as were a chest X-ray and an ultrasound of her abdomen and pelvis. A spinal tap was also unrevealing. Despite the normal test results, the team started the infant on antibiotics just in case. But after three days, the baby still wasn't eating. She was too weak to lift her head. That's when the parents asked that their baby be transferred to a nearby hospital that specialized in caring for children.

That evening, the baby was sent to the Morgan Stanley Children's Hospital in Manhattan. Dr. Pelton Phinizy, a pediatrician in his last year of training, was the resident on call. When he saw the ambulance crew wheeling a bassinet into the ward, he followed. He scanned the thin file sent from the community hospital, then went in to meet the parents and the child.

As soon as he saw the baby, Phinizy sensed something was seriously wrong. “All my alarms were going,” the third-year resident told me. The baby lay splayed on the bed like a starfish, completely motionless. Her eyelids were droopy, as if she were constantly on the brink of sleep. But the doctor worried most about the quiet gurgle coming from the baby’s throat.

The parents told the young resident about the baby’s strange refusal to eat. Until a few days before, she was perfectly healthy. She hadn’t been injured or exposed to anyone who was sick. She and her older sister went to the park most days, but the older child was fine, with no symptoms at all.

Since becoming sick, the baby had not run a fever, but her parents suspected she might be congested—she’d been making that odd gurgling sound for the past couple of days. And she was a lot crankier than usual, as if she wasn’t feeling well.

Once he spoke with the parents, Phinizy examined the baby. She was plump, and her eyes were surprisingly alert behind their half-closed lids. But when the doctor picked her up, her head flopped back as if it were too heavy for her neck to support. Her arms and legs hung straight down, and she made no effort to move them. Carefully, he laid her back onto the bed. Picking up one arm, he pulled it across her chest. Normally the shoulder muscles would prevent an infant’s hand from reaching past the opposite shoulder. Her arm felt like dead weight, and her hand stretched well beyond the shoulder. It lay strangely flat against her chest, like a scarf pulled too tight. When he touched the back of her throat with a tongue depressor, she gagged, but the rest of her reflexes—in her arms and legs—were completely absent.

Phinizy went back to the records from the first hospital. Doctors there had already ruled out some of the most likely causes. The tests on blood, urine, and spinal fluid suggested that the illness was probably not infectious. The ultrasound of the baby's intestines showed that there was no obstruction or displacement. So what was going on?

Could this be the first sign of some congenital disease of the muscles or nerves? The family had nothing in its history that would suggest such a thing, but some of these diseases require that both parents pass on a defective gene in order for it to develop. Was this Guillain-Barré syndrome? That paralytic disorder can cause the kind of widespread muscle weakness this baby displayed. Could it be an infection in the brain that went undetected in the spinal tap? Early encephalitis can be missed this way. What about botulism? It's very rare—there are usually fewer than 150 cases a year in the United States—but it can cause this sort of profound and sometimes deadly weakness.

If there was one thing Phinizy was sure of, it was that this baby was much too sick to be cared for in the usual pediatric ward. The gurgling sound in her throat worried him. It suggested she was too weak to swallow. She could drown in her own saliva, as she nearly did just a few days before at the community hospital. She really needed to be in the intensive care unit to be monitored and, if necessary, put on a ventilator. So Phinizy headed to the ICU and tracked down Dr. Stanley Hum, the pediatric intensivist on call that day.

Phinizy outlined the case and described his physical exam. Hum was particularly intrigued by the suddenness of the paralysis in an otherwise healthy child—few diseases cause that. Guillain-Barré could, but that should have shown up in the spinal fluid. Even without seeing the baby, and having encountered the disease only once before, he recognized a classic presentation of infant botulism.

Botulism is a rare, potentially deadly disease caused by a nerve toxin made by the bacterium *Clostridium botulinum*. Muscles exposed to the toxin become paralyzed, losing the ability to contract. If the paralysis affects the diaphragm, sufferers can asphyxiate if they don't get medical help.

The disease was first described in the nineteenth century, when hundreds in Germany were paralyzed after eating sausage that had been contaminated with the bacterium and its potent toxin. "Botulinum" and "botulism" come from the Latin word for sausage, *botulus*.

Although botulism was first described as a food-borne illness, most cases come from exposure to soil contaminated with the bacterium. Infants make up most of the cases worldwide. Their incompletely colonized colons make it easy for the bacterium to get a toehold and start making its poison. The baby may have been exposed to the bacterium in the soil from the park. The recent flooding caused by Hurricane Sandy may have deposited it anew.

The definitive tests for botulism take days, and treatment is most effective when started early. So even before confirming the diagnosis, Hum told the resident the baby would need the botulism antitoxin, called BIG-IV, which provides the baby with antibodies to help stop the disease's progress. Phinizy hurried back to tell the parents. Hum and his team contacted the makers of BIG-IV in California, and it arrived the following day.

Not long after moving to the ICU, the child started to have trouble breathing, so Hum put her on a ventilator. The diagnosis of botulism was confirmed days later. The baby remained in the ICU for two and a half weeks while she recovered. She was finally released from the hospital after about a month, not quite as plump as before, but back to her usual happy, hungry self.

In a recent study of diagnostic errors, Dr. Hardeep Singh noted that more than three-fourths of the errors occurred in the initial meeting between doctor and patient, and that most of them were a result of inadequate history gathering or an inadequate physical exam. In this case, Dr. Phinizy was able to get a detailed history from the child's parents and from the note sent from the first hospital the baby went to. He also took the time to do a thorough physical exam. He wasn't sure what the baby had, but his careful data collection allowed a more experienced doctor to make the diagnosis without even seeing the child.

Our high-tech diagnostic tools get all the glory in medicine. But it turns out that most often it is the old-fashioned skills—listening to the patient, examining their body—that allows us to make the right diagnosis.

Total Collapse

"Can you help me?" The fifty-two-year-old father called out to his son, who was asleep in the other bedroom. It was nearly midnight, and the man, awakened from sleep, had tried to get up to go to the bathroom. When he stood, he was surprised to find that his legs had no strength, and he fell. Now he needed help to get back up. His son, twenty-one and mentally disabled, came into the room. His father quietly talked him through what he had to do to get him onto the bed. Then he picked up the phone and called a friend to come stay with his son. Once that was arranged, he dialed 911.

Dr. Kathleen Samuels, the resident admitting patients to the intensive care unit that night at Waterbury Hospital in Connecticut, had already admitted several by the time she got the call about the man who couldn't walk. He had a life-threateningly low level of potassium in his blood. Potassium is an essential electrolyte, and the body normally holds it at a constant level. The ER doctor hadn't figured out why this man's potassium was so low, but he was certain the man needed to be monitored in the intensive care unit until they understood what was going on.

Samuels hurried to see the patient. He appeared healthy and seemed surprised that he couldn't walk. He told her that he was well until two days earlier, when he started having pain in his hips and knees. The pain was constant but was worst in the morning and when he walked. He went to the emergency room twice in those two days. The first time, he left after a few hours. He hadn't seen a doctor, but he had to pick up his son from his daycare program. He came back the next day, and the ER doctor told him it was arthritis and prescribed a painkiller. He didn't get a chance to pick up the medicine, and now he couldn't walk. He had no other medical problems, and he took no medicines. He didn't smoke or drink, and he took care of his two disabled adult children—one of whom had been in the hospital for the past two weeks.

The patient could lift his legs off the bed, but he wasn't able to keep them up if Samuels applied even a little pressure. He had a slight tremor, but the patient told her that he'd had that for years. Otherwise the exam was unremarkable. His labs, on the other hand, were anything but. Not only was his potassium low but so, too, were his white-blood-cell count and platelets. His blood sugar was high and so was his thyroid hormone. The thyroid gland tells the body how hard to work, and this thyroid was telling the body to work very hard indeed. But it was only the very low potassium that could kill the man, and that's what commanded Samuels's attention. She made sure that he was given enough potassium to replace what he'd lost and then tried to figure out why he lost it.

Diarrhea and vomiting are common causes of low potassium. This patient said he had neither. Potassium is regulated by the kidneys. Although his kidneys appeared normal, he would need more extensive testing. Some medications cause the kidneys to dump excessive amounts of potassium into the urine, but this patient took no medications.

At 7:30 that morning, Samuels went to the hospital's Resident Report, a daily meeting for physicians in training, where much of the teaching on diagnostic thinking takes place. Residents and teaching physicians gather in a conference room to think through the process of making a diagnosis for a patient admitted to the hospital. This morning Samuels laid out the case of the man who couldn't walk—how the patient looked, what she found during the exam, and what his lab results revealed.

As the doctors hashed out the case, Dr. Jeremy Schwartz, one of the chief residents, remembered something. This patient's symptoms sounded just like an illness he'd read about: a genetic disorder called hypokalemic periodic paralysis. "Hypo" from the Greek, meaning "low," and "kalium" from the Latin, meaning "potassium." In this disease, patients experience transient episodes of severe weakness caused by low potassium. But there was one big difference—this inherited disease is usually first seen in adolescence. This man was much too old to be having his first attack. Could there be a form of the disease that was acquired and not inborn? Could it be linked to something else the man had—to his high thyroid or sugar?

Schwartz was sitting next to the computer in the conference room. He went to a medical reference site and typed in "hypokalemic periodic paralysis" and "hyperthyroidism." As soon as he hit enter, page after page appeared, filled with articles on a disease known as thyrotoxic periodic paralysis.

In the inherited version of hypokalemic periodic paralysis, young men (mostly) are born with cells that can suck up potassium after a high-carbohydrate meal, after exercise, upon awakening from sleep, or during times of intense stress. Patients with this genetic disorder can reduce their risk of paralytic attacks by taking medicines that increase the amount of potassium in the blood and by eating a low-carbohydrate diet.

This man didn't have this genetic disease, but having too much thyroid hormone made his body act as if he did. When high levels of thyroid combine with high blood sugar, a high-carbohydrate diet, or high stress, cells can take up so much potassium that there's just not enough outside the cell, where it's needed for muscles to work. This patient had it all. Blood tests already showed that he had high levels of thyroid hormone and, at the time of his arrival to the hospital, a very high blood sugar. He was experiencing high levels of stress because his older son was in the hospital, and in addition, he was living on high-carbohydrate foods from the vending machines there.

Still, hyperthyroidism is common; high blood sugar is common; high-carbohydrate diets and stress are epidemic; and yet this kind of periodic paralysis is rare. Current thinking is that these patients also have a genetic abnormality that predisposes them to develop periodic paralysis if and when they ever develop hyperthyroidism.

The patient was given a small dose of replacement potassium by mouth, and his potassium and his strength returned to normal. He was started on a thyroid medication, which alleviated the pain and weakness.

I heard about this patient because I was his primary care doctor. I last saw him two years before this episode, when he made an appointment because he had heartburn. I gave him a medication for the heartburn, but I also noticed that he had a rapid heart rate and a tremor, and I suspected that he had hyperthyroidism. I gave him a lab slip to check his thyroid hormone. He never went. And I didn't have a system then to follow up on patients to make sure they get the studies I prescribe. I do now.

When I saw him after he was discharged from the hospital, I asked why he had never gotten the blood test. He looked a little embarrassed, but his answer was direct: His complaint had been the heartburn, and the pill I prescribed fixed that. Concerns about hyperthyroidism didn't seem important to him. He didn't care about a body part he had never heard of, possibly causing a disease with symptoms he didn't feel.

All that changed when he lost his strength. Suddenly, he told me, he was quite literally unable to care for his sons. "If I am not around, they will have no one," he told me. So now he takes his thyroid medicine regularly. He gets his blood drawn as often as needed to keep his disease in check.

"I don't do it for myself," he said. "I have to take care of myself so I can take care of my kids."

Fear of Falling

"What can you do for me that all the doctors who have already seen me haven't?" the woman demanded. Her face was puckered with frustration, her voice edged with irritation. Poorly fitting dentures clipped her words. "I'm too weak to walk and almost too tired to care," she added, her voice dropping to a whisper. Dr. Bilal Ahmed nodded sympathetically. He had heard about the woman's mysterious debility from the resident who admitted her to Highland Hospital in Rochester, New York, the night before.

A couple of years earlier she started "walking like a drunk," she told the slender, middle-aged doctor. Her legs were weak and her feet were numb. The only feeling she had in them was a pins-and-needles sensation, as if her feet had gone to sleep and never woken up. A few months ago she started falling. She broke her ankle in a particularly bad fall; the ankle got better, but she didn't. Now she was in a wheelchair.

Her internists referred her to a neurologist, who sent her to the hospital for an MRI. After the test she was so weak that the doctors were reluctant to send her home, and she was admitted to the hospital. And here she was, hoping for an answer.

Ahmed was surprised to learn that she was only sixty-four. Her face was deeply lined, and her eyes were puffy and dulled with fatigue. Her long hair was dark but peppered with gray. Her shoulders and arms had normal strength, but her legs were weak; she could lift them off the bed but couldn't keep them up when the doctor tested their strength. And she had lost most of her ability to feel hot or cold or even a light touch from her feet to her knees. When the doctor helped her stand, her legs were strong enough to hold her up, but she wobbled dangerously when he let her go. He helped her back in bed before she could fall.

She had diabetes and high blood pressure. And she was seeing a hematologist for a dangerously low white-blood-cell count that still was not explained. But none of that helped account for why she was so unsteady on her feet. Although her legs were weak, Ahmed thought that it was mostly her balance that kept her from walking. Balance is a function of the nervous system. Diabetes can damage the nerves of the legs and feet but rarely causes this degree of disability.

Some type of injury to the spinal cord could cause this loss of feeling and coordination. Had the narrow, bony tube that protects the spinal column been made even narrower by an overgrowth of bone? That kind of pressure on the nerves is a relatively common problem as people get older. Some cancers could do this as well. A wildly proliferating tumor will sometimes create proteins that attack nerve cells. And cancer could account for her low white-blood-cell count as well. Were the two symptoms linked? A deficiency of vitamin B_{12} could affect the nervous system and the production of white blood cells. And it was also common in people over sixty. The patient remembered being bitten by a mosquito before her symptoms started. Could this be the West Nile virus? It often attacks the spinal cord, causing weakness that can be permanent.

Ahmed quickly came up with a plan. Blood tests would show if she had been exposed to the West Nile virus or if she was low on vitamin B_{12}. He would also test her for a cancer of the blood that could affect both blood and bone. The MRI could tell him if the spinal cord was damaged or if she had cancer.

After finishing his notes, Ahmed went to the radiology department to check out the MRI of the patient's upper body and spine. It was normal. There was no sign of a cancer in the chest, abdomen, or pelvis, and no sign of any damage to the spinal cord.

The next morning, results of the blood tests began to roll in. The vitamin B_{12} level was normal. She had never been exposed to the West Nile virus. There was no evidence of any blood cancer, either. Ahmed saw that the neurologist had ordered some blood tests before sending the patient for the MRI—tests that Ahmed said he would not have thought of running. Two were strikingly abnormal: she had almost no copper in her system; and her zinc level was through the roof—more than twice the normal amount. Ahmed was surprised. Could abnormalities in these minerals do all this? He hurried to a computer to read up.

Tiny amounts of copper are needed to drive a handful of essential cell functions. Without copper, nerve cells in the spinal cord aren't able to do their work of providing data from the world outside the body to the brain; eventually they will die. Copper deficiency can also cause a low white-blood-cell count. But it is rare, mostly because we need so little copper and it is so abundant in the foods we eat. So why wasn't this woman getting enough? He found the answer just a few pages later. It was the high zinc level in her system. Although zinc is also essential for normal cell function, too much will cause the body to dump copper. She had too little copper because she had too much zinc. Okay, but why did she have too much zinc?

Ahmed returned to the patient's room with some answers and a few more questions. He explained that her weakness and loss of balance were probably due to a copper deficiency; she would need to take copper supplements for the next several weeks. But she also had too much zinc in her system, and they had to figure out why. Did she take zinc supplements or use zinc cold remedies? No, she told him, never. She had worked in several factories, but that was at least twenty years ago. Contaminated well water can contain high levels of zinc, but she used city water. So where was all this zinc coming from?

On the bedside table, Ahmed noticed a half-empty tube of denture adhesive. He picked it up. Didn't some of these adhesives contain zinc?

"Leave that alone!" the patient shouted, suddenly enraged. "That is my only comfort. It's the only way I can eat, and now you want to take that away from me? No. Put it down." The fear and frustration were clear on her face.

Ahmed had noticed that her dentures didn't fit well, even with the adhesive. It's a common problem. The jawbone recedes after teeth are removed, and dentures often need to be replaced every few years to maintain a good fit. Ahmed asked how much denture cream she used. Oh, a lot, she told him. She went through a tube every day or so. She might use five or six tubes a week.

That was it. When used as instructed, a tube of adhesive should last for a month or more. This woman's dentures didn't fit well, so she needed more than the recommended amount to hold them in place. Much more. And she had been doing this for years.

I spoke with the patient a year and a half after Ahmed figured out what the problem was. While she hadn't been able to afford new dentures, she was using a denture adhesive that didn't contain zinc. Her blood count was back to normal and she felt better. She no longer had the nagging fatigue that plagued her before she went into the hospital, but she still couldn't walk without a lot of help. She continued to go to physical therapy, but the damage to her nerves may be permanent.

Ahmed was encouraged by the small amount of progress his patient had made but said he worried that others might be suffering from the same problem. "There are only three published reports of this kind of toxicity from denture adhesive," he said. "Does that mean it's rare, or just rarely diagnosed? All I know is that now I look for it and I didn't before."

An Overwhelming Weakness

The white-haired man looked up from his newspaper as the doctor entered the hospital room. His eyes were a bright blue, and a warm, somewhat crooked smile lit up his face. "Sorry I can't stand," he said gallantly after the doctor introduced herself. "My legs are weak." Dr. Merceditas Villanueva, a specialist in infectious diseases, returned the smile, then asked the patient to tell her about the weakness.

He was seventy-seven. Never sick a day in his life—until the week before, on the Fourth of July. That day, also his wife's birthday, the patient's children and grandchildren had come to spend the day at his pool. In the late afternoon, the patient tried to get up to start the family dinner. "I do most of the cooking, especially on the holidays," he told her in his lightly accented voice. "I'm Hungarian, and I like to cook the foods from home." But that afternoon he was surprised to find that getting out of his chair was strangely difficult. He struggled to his feet, shrugging off the sons who hurried to help, and made his way slowly to the kitchen. But he hadn't been able to cook that night—or any night since then. By the end of the week, he had no strength at all. "I couldn't even take a step," he said. "I was helpless." His face eased into his lopsided smile once more. "My wife insisted I come to the hospital."

In the meantime, he also developed pain and swelling in his right knee, his left elbow, and both feet. But, the patient added, joint pain was nothing new. The weakness, though, had never happened before; that was a little worrisome.

The patient didn't smoke and drank a glass or two of wine with his dinner most nights. His only medical problem was high blood pressure, which was well controlled with daily medication.

In the emergency room, he had had a fever of 102. His right knee was red, warm, and markedly swollen—the tender flesh as squishy as a water balloon. His right foot, left big toe, and left elbow were also inflamed. The strength in his arms and shoulders was normal, but his legs were so weak that he could barely lift them off the bed.

Neurology was consulted. The loss of strength in both legs suggested that the problem wasn't in the brain (where left and right are kept separate) but was located either in the spine or in the nerves of the limbs themselves. To distinguish between these two possibilities, the doctor then tested another aspect of the nervous system: sensation.

Using a pointed probe, he touched the patient lightly on each leg, making his way from toes to thighs. The patient assured him at each step that he could feel the light pinch of the probe equally on both legs—but only barely. The neurologist moved up to the abdomen. Just past the navel, the patient suddenly felt the probe sharply. One of the great beauties and pleasures of neurology is that the well-trained physician can often pinpoint the precise location of a problem within the vast and complex nervous system. Pinpointing the spot where the patient's sense of feeling changed from dull to sharp told the neurologist, first, that the lesion involved the spinal cord and, second, where on the spine it was located.

An MRI confirmed what the doctor surmised: near the bottom edge of the rib cage, within the spinal column, there was a patch of dark where there should be light. This suggested a fluid collection inside the bony spine but outside the tough sac that contains the spinal cord, in an area called the epidural space. The fluid was compressing the spinal cord and causing the weakness and loss of sensation.

The most common cause of this kind of fluid collection is an infection, an abscess. Untreated, an epidural abscess can progress rapidly, causing paralysis and, rarely, death. The patient was given potent intravenous antibiotics, and Villanueva was called.

After hearing the patient's story, the doctor examined him. He still had a fever, and his knee remained swollen and painful. But Villanueva also took note of something either not observed or not thought important enough to have been mentioned: Not only was his left elbow red and swollen; the joint itself was grossly deformed by several large, firm, irregularly shaped nodules that she immediately recognized as tophi—the crystalline residue of severe gout. Once known as the "disease of kings" because of its association with rich foods and wine, gout is the result of a buildup of waste products in the body that intermittently crystallize in the joints and cause characteristic bouts of pain, swelling, and inflammation. Although the disease is common, the crystal deposits so obvious in this patient's elbows are an unusual finding these days. Effective medicines now prevent the recurrent attacks of inflammation and crystal deposits that form these strange firm nodules.

This patient appeared to have severe gout, the doctor mused. That was the likely cause of the joint pain, but what about the weakness? It seemed a bit far-fetched. She had never heard of gout traveling to the inner part of the spine, where this fluid collection was located.

The medical team worried that the pain and weakness could be caused by an infection, and Villanueva agreed that that diagnosis made a certain sense. An epidural abscess usually occurs when an infection somewhere else in the body—in this case the joints—spreads to the spine through the bloodstream. But blood cultures taken the day of admission had not grown any bacteria. Moreover, to Villanueva's experienced eye, this patient didn't appear sick enough to have a widespread infection.

Gout can mimic infection. Fever, red-hot joints, and an elevated white-blood-cell count are also characteristic of a gouty attack. Which was it—infection or gout? Or was it both?

Villanueva needed to find out whether there was bacteria in the fluid in his knee. If so, that bacteria could have spread to the spine. The final lab report showed no bacteria, only the needlelike crystals of gout. The patient was started on anti-gout therapy.

Could the fluid in the spine also be caused by his gout? Was that possible? Villanueva sat down in front of her computer to search the medical literature. Finally she found two case reports of gout invading the spinal column. It was rare, but it could happen. Before she could stop the antibiotics, though, she would have to prove that had happened here.

In another era, a doctor might have been willing to just treat for both diseases. The patient was doing quite well on this dual therapy: His fever was gone; his weakness was resolving. He was even walking—albeit with assistance. Nevertheless, Villanueva was reluctant to consign the patient to up to six weeks of antibiotics if it wasn't necessary. Overuse of these drugs contributes to the development of superbugs—bacteria resistant to all medications. No, the patient needed a definitive diagnosis; she needed to know what was in the fluid in his spine.

The next day, a radiologist carefully inserted a needle into the tiny collection of fluid and drew out a few cc's of blood-tinged fluid. Under the microscope, the diagnosis was confirmed—there were crystals but no bacteria.

I visited the patient a few weeks after he went home. It turned out that he had never told his doctor about his painful flare-ups and therefore had never been given medication for them. Without it, each attack of gout had become progressively worse, until it ended up in his spine. His gout, then, was a throwback to an earlier era. And like those kings of old, he had been nearly crippled by it. His walking was still slow but getting better daily. He wasn't cooking yet, but he planned to start soon. Very soon.

The Long Haul

"I guess it's my time," the fifty-year-old man said quietly as his wife drove him home from his doctor's office in southwest Houston. "No," his wife responded sharply. "I won't accept that. We're going to get a second opinion." To the patient, a software engineer and father of two teenage girls, it seemed as if this was the end of his fourteen months of rapid decline from a robust man to this skeleton who needed a walker to get from car to door.

Just two summers before, he was healthy. By fall, he noticed that his morning walks with his golden retriever were getting shorter. His thirty-minute route became twenty, then ten, then even less. His feet felt like bricks: cold, unfeeling, heavy. A simple trip to the park or upstairs to their bedroom felt like a workout.

First, he went to see his primary care doctor. He had no fevers or chills. His appetite was good, though he'd lost weight. He was sleeping fine. His exam seemed normal until the doctor tapped the front of his knee with his little rubber hammer. Nothing happened. There was no kick reflex. He tried again; still nothing. He tapped the tendons behind his ankles and the ligaments in his forearms, but he couldn't elicit the normal jerks of an intact reflex.

His doctor wasn't sure what was wrong, but he was worried. He referred his patient to a neurologist for further evaluation. When that doctor examined him, he noticed significant weakness in the man's hip and leg muscles. His shoulders and arms were not as strong as they should have been. And he had no reflexes anywhere. The neurologist reviewed the blood test obtained by the primary care doctor. It wasn't his muscles. It wasn't his thyroid. There was no sign of anemia, infection, or inflammation.

The loss of reflexes indicated that the man's weakness was probably caused by his nerves rather than his muscles. The neurologist performed a nerve-conduction study, during which a tiny electrode, inserted into the muscle, measures the electrical impulses the nerve fibers send to the brain when the muscle is working. In areas of the patient's weakness, the test showed that the fine strands of nerve tissue were damaged and transmission was slowed. A spinal tap revealed high levels of protein in the cerebral spinal fluid, suggesting that the patient might have some type of autoimmune neuropathy, nerve damage caused by rogue antibodies. The neurologist thought the patient most likely had either Guillain-Barré syndrome (GBS) or its longer-lasting cousin, chronic inflammatory demyelinating polyneuropathy (CIDP).

The doctor first treated him with plasmapheresis—a technique that removes antibodies from the circulation. This process can be used to treat both GBS and CIDP. It worked. In the weeks following that first series of treatments, the man felt stronger. He was excited to be back to about 70 percent of normal. But it didn't last. Two months after his last treatment, he started losing ground. His balance went, and after a couple of falls, he had to start relying on a cane. The doctor changed his treatment to IVIG—intravenous gamma globulin, flooding his system with other people's antibodies. Maybe it helped a little, but he continued to weaken. The cane stopped being enough, and he had to use a wheelchair outside the house.

Because the patient was not responding to treatment as expected, the neurologist looked for other possible causes of the severe weakness. Was it HIV? Could it be lupus? Or a rare cancer of white blood cells called POEMS? He sent out more blood and scanned his bones, but those results revealed nothing new. Imaging showed no pathology in his brain or spine to account for the weakness. He put the patient back on plasmapheresis and sent him to physical therapy. He continued to worsen. His hands became clumsy, and eating was difficult. He also kept losing weight. He started out at just over 210 pounds. After that year of treatment, he was down to 150.

Throughout this devastating illness, the man's wife had taken the research and problem-solving skills she'd developed as a lawyer and tried to apply them to his case. She recalled that the neurologist had mentioned the possibility that he had POEMS—an acronym of the symptoms caused by an overproduction of antibodies, including polyneuropathy (pain and weakness and a loss of sensation in different parts of the body), organomegaly (enlarged organs), endocrinopathy (hormonal abnormalities), monoclonal plasma cell proliferation (too many antibody-producing cells), and skin changes. But he had ruled out that diagnosis.

Reading up on it, she found case reports of patients who

sounded remarkably like her husband. He had at least two of the symptoms (polyneuropathy and monoclonal plasma cell proliferation). Maybe it is POEMS, she suggested to the neurologist. Not likely, he said. It's quite rare, and her husband didn't fully fit the diagnostic criteria. Besides, the proliferation of antibodies that made him think of POEMS in the first place is seen in up to 10 percent of patients with CIDP. He cited the aphorism leveled at all who suspect "zebras": An unusual manifestation of a common disease is much more likely than even a classic presentation of a rare disease.

That's when the patient's wife decided they needed a second opinion and made an appointment with Dr. Kazim Sheikh, a neurologist at the University of Texas at Houston. The first neurologist had encouraged them to seek a new neurologist, because if the patient didn't have CIDP, he wasn't sure what he had.

She took her husband to see Dr. Sheikh a few weeks later. He listened to the patient and his wife describe their hellish year, and even before he examined the patient, he made his first diagnosis: "I think you have a cancer. I'm not sure what kind, but I think you should be seen at MD Anderson Cancer Center." The patient's wife shared her thoughts about POEMS syndrome. It might fit, the doctor agreed. But there were also other possibilities. He ordered a PET scan to look for lesions on the bone. In POEMS, patients often develop tumors made up of cells that produce antibodies. A PET scan looks for cell activity, and because malignant tumors are constantly producing abnormal cells, they light up on this kind of scan. There, on his pelvic bone, was a tiny hot spot, a tumor, too small to be seen on the earlier scan.

The patient was referred to MD Anderson, where he was finally given the diagnosis of POEMS. Initial treatments were ineffective. His last chance was a stem-cell transplant. This is a difficult therapy in which, after storing a sample of stem cells that can grow into both red and white blood cells, the patient is blasted with chemotherapies designed to kill everything left behind. Only then can the harvested cells be reintroduced to their newly cleaned bone marrow to repopulate. Long-term remission rates after treatment are good—98 percent after one year and 75 percent after five.

At this point, though, the patient's doctors were concerned that he was too sick to survive the stem-cell transplant and chemotherapy. His wife insisted that the doctors should let him take the risk of dying from the treatment rather than be certain of death without it. Seeing the patient, Dr. Muzaffar Qazilbash, the transplant oncologist, was reminded of Stephen Hawking, with his six-foot frame folded into a wheelchair, shrunk down to just over a hundred pounds. He reevaluated the man—could he survive the procedure?—and agreed.

The patient made it through the transplant and then the many complications afterward. Slowly, slowly, he recovered. Four years after his transplant, he no longer needed a cane and was back to walking. He was once again driving his daughters to their after-school events. And he was back to work, full time.

He knows that he is one lucky man. He was lucky to have the insurance he had, lucky to have survived both the disease and the treatment, and really lucky to have a wife who wasn't going to let him go.

Wasting Away

"He wasn't always like this, you know," the woman said. She turned her tanned, earnest face away from her son, still strapped in his wheelchair, and stood to face the doctor. "He was a pretty normal kid." The physician, Dr. Joel Ehrenkranz, focused his dark, intent eyes on the wheelchair and its motionless passenger. His long, thin legs were tucked tightly against the chair; his chin rested on his chest as if his neck were too weak to hold it upright. Raised pink scars slashed across his scalp—evidence of some long-ago surgery. His face was thin and drawn; his eyes were dull and inattentive. A pale fringe of light hair barely covered his ears.

At thirteen, he found out he had a brain tumor. He survived an operation and an infection. Then he got better. With physical therapy, he learned to walk again, and he eventually went back to school, back to his friends, back to his life. Then he had more operations, followed by radiation, and his health began to unravel. "That was twenty-five years ago," his mother said. "Now he's just a ghost of who he used to be." At forty-three, the patient hadn't walked in more than five years. He had hardly spoken in a decade. "We initially thought he'd recover completely, but somehow he's only gotten worse." Over the years, he had seen many doctors—surgeons, neurologists, endocrinologists. One started him on steroids, and that helped for a while. Thyroid hormone helped, too. But nothing had stopped this progressive decline, or even really explained it.

Then, just over a year ago, his mother took him to the community hospital for an operation on his legs. Years of immobility had shortened and contracted his muscles and tendons. The operation was supposed to fix that and make him more comfortable and flexible. "But after the surgery, he really lost it," his mother reported. A couple of days later, he started vomiting everything he was fed. Every touch seemed painful. "He didn't know what day it was or even where he was. I don't know if he even knew who I was. You could put a ringing phone in his lap and he wouldn't respond at all." She wondered if the end had finally come.

Two months after that surgery, she took her son back to their family doctor, David Sherwood, she told Ehrenkranz. As one of just three physicians attending to the residents of their Colorado town, he had known and cared for the young man and his mother for the past couple of years. He was shocked by the change after the operation. "I'd seen him quite a few times," Sherwood recounted to me. "He'd always been about the same—debilitated and weak, but alert. You felt that even if he couldn't express himself, he could understand. He had a tough history with the brain surgery and all, and I figured that he was who he was because of that. Seeing him after this fairly minor procedure, it was like all the life was gone from him. I got this gut feeling I was really missing something here." The patient had already been to the local specialists, so Sherwood was reluctant to send him back. Instead, he suggested that they drive across the state to see Dr. Ehrenkranz, an endocrinologist who had recently come to his attention.

As she finished her story, the patient's mother looked expectantly at the doctor she had traveled so far to see. Ehrenkranz said nothing throughout her long and complicated tale. Occasionally he glanced through the voluminous records she had brought from the years of their ordeal. Mostly he looked at her son.

At last, Ehrenkranz reached into an old-fashioned black medical bag and took out his stethoscope. He gently wrapped the blood pressure cuff around the patient's wasted arm. His pressure was remarkably low, 90 over 70. A normal blood pressure for a man in his forties was maybe 120 over 80. "That was a clue," Ehrenkranz recalled. "His tumor, all the surgeries on his brain, that wouldn't affect his blood pressure." He had almost no hair on his face or body. The incisions in his groin from his recent surgery were well healed, without any redness or swelling.

Ehrenkranz focused on the most striking physical finding: the low blood pressure. Infection was the most common cause of such a symptom. The patient didn't appear acutely ill, the way he would if he had a raging infection. But could he have an infection at the surgical site? An abscess—a walled-off infection—was capable of this type of insidious process. Still, the patient had not shown any evidence of pain when the doctor touched his surgical scars. Ehrenkranz had felt no lumps or bumps suggestive of a well-contained pocket of pus. Low blood pressure could also be caused by a heart that's not pumping properly. Yet the physical exam suggested that his heart was a normal size, and there were no extra heart sounds that would suggest a faulty valve.

Either an infection or heart disease was possible, and yet neither quite fit. Ehrenkranz continued to look at the patient. "It's like figuring out a math problem," he told me. "You just keep looking and looking, and it's like a light goes on." What if the patient were in a physiologic crisis after this operation because he lacked an essential hormone: cortisol? He was on a low dose of a steroid hormone, and while it might have been enough to keep him alive, it wasn't enough to deal with the stress of surgery. The nausea, vomiting, and loss of alertness could be symptoms of insufficient stress hormone. But would this have caused his slow deterioration, too? What if he were missing more than just stress hormone? He had also been put on thyroid hormone. Both are controlled through the tiny gland at the base of the brain called the pituitary. Referred to historically as the "master gland," this complex structure regulates the production of thyroid hormone and cortisol, the stress hormone. It also regulates growth hormone and sex hormones, including testosterone. The doctor theorized that this master gland had been destroyed by the radiation the patient had received decades earlier, and that his subsequent decline was driven not by the brain destroyed by the tumor and its surgical cure but by the loss of the pituitary.

At once, Ehrenkranz knew he was right. He sent the patient to the lab to have blood drawn to confirm his suspicion but was confident enough to start treatment. He wrote a prescription for large doses of steroid hormones to replace those the patient could no longer produce himself. He also wrote prescriptions for growth hormone, thyroid hormone, and testosterone. Two days later, the patient got his first "stress dose" of steroids. The next morning, the mother went to see her son. He looked up at her, said, "Hello, Mom," and smiled a tiny smile. They were his first spontaneous words since the groin surgery. Ehrenkranz got the test results confirming the diagnosis the next week, but the patient's mother already knew. "I didn't need any tests to tell me the doctor was right," she said. "Two days after starting on these hormones, my son was able to stand up just using handrails. He hasn't been able to do that since the surgery."

Ten months later, he was talking, eating, and working out. He had gained about forty pounds and grown a beard. He was listening to music, drawing, telling jokes. He had even been out riding horses—a childhood passion. Ehrenkranz calls him Rip Van Winkle, just waking from a long, long sleep. He was still using a wheelchair—his legs and body remained weak from the decades of debilitation. I spoke with him just before Thanksgiving that year. He said he had a lot to be grateful for. And he was certain he would walk one day. "Someday soon," he told me.

Missed Signals

The emergency technicians burst through the doors, pushing a stretcher into the crowded ER. Their walkie-talkies dangled from their shoulders, squawking and hissing like demented parrots. The triage nurse directed them straight into a room as the EMTs barked out what they knew. "Sixty-four-year-old man . . . history of a stroke . . . complaints of weakness and belly pain." His heart was slow, they reported; his blood pressure so low that it was not measurable. The monitor showed a heart rate in the twenties—normal is over sixty. Dr. Bernd Woerner strode in and quickly assessed the situation. "Get me an amp of atropine," he snapped, calling for the medicine used to speed up the heart.

The doctor watched as the monitor screen continued its flat yellow line, broken only occasionally by the spike indicating another heartbeat. Slowly the patient's heart rate and blood pressure began to rise.

Throughout all this the patient was alert, Woerner told

me later. He explained to the patient, "Your heart is pumping too slowly." The medicine would keep his heart rate up until the cardiologist arrived in an hour or so to insert a pacemaker. In the meantime, they had to begin to figure out what was wrong with his heart.

I knew this patient. I was his internist and had been seeing him for the past year, since he had his stroke. Before that, he hadn't been to a doctor for decades. He came to me when a massive stroke rendered his right leg and arm nearly motionless, his face crooked and his speech slurred. Still, his beautiful cockeyed smile and gallant manner made him a favorite at our office. He often brought us gifts—candy or some of the pecans sent from his family in North Carolina. He was doing well, so I was shocked when I got word from the ER that my patient was dying. And the doctors there weren't sure why.

With the usual chaos of the emergency room boiling around them, Woerner forced himself to sit quietly as the patient described his symptoms. The man spoke in an unnaturally deliberate drawl, as if in slow motion: "I—can't—walk." It started the night before. He felt weak, could barely move. Any chest pain? Woerner broke in. Shortness of breath? Fever or chills? Vomiting? The patient shook his head no. He was taking medications to lower his blood pressure and cholesterol. He had not smoked or drunk alcohol since his stroke. Examining him, Woerner saw the results of the stroke but little more.

Why was his heart beating so slowly? the doctor wondered. Had he taken too much of one of his medications? Had he suffered a heart attack that affected the natural pacemaker in his heart?

Part of the answer came less than an hour later. The lab

called to report that the patient's kidneys weren't working. And his potassium—an essential element in body chemistry, regulated by the kidneys—was dangerously high. Potassium controls how easily a cell responds to the body's commands. Too little potassium, and the cells overreact to any stimulation; too much, and the body slows down. The patient was given a medicine to get the potassium out of his system and then transferred to the ICU for monitoring.

If the potassium was high because of his kidney failure, what had caused his kidneys to fail? Dr. Perry Smith, the intern on call in the ICU, gnawed at this question as he reviewed the chart and examined the patient. It wasn't a drug error. The patient's medication box showed the correct number of pills. And it hadn't been a heart attack; a blood test proved that. Smith looked for the results of the urinalysis to see if there was any clue there. Somehow no one had sent any urine to the lab. Were his kidneys too damaged to produce urine? That would be important to know. Smith asked the nurse to get some urine from the patient.

She returned empty-handed. The patient couldn't urinate, and she hadn't been able to insert a Foley catheter, a rubber tube that is passed through the urethra into the bladder to collect urine. Was something blocking the urethra? A urology resident finally managed to get a catheter into the bladder. Urine gushed out—nearly half a gallon of it. A full bladder normally holds only a quarter of that. The urology resident looked at the intern: "I guess now we know why his kidneys weren't working."

The urethra was blocked—by the prostate gland. The prostate surrounds the urethra, and when it enlarges, as it often does with age, it can impinge on the narrow outlet, obstructing and ultimately blocking it so that no urine can pass. As the trapped liquid filled the bladder, the pressure shut down the patient's kidneys.

Just hours after the obstruction was relieved, his potassium began to drop as the kidneys went back to work. Four hours later, the patient's heart rate was up over sixty. By the next morning, the abdominal pain, probably caused by his hugely distended bladder, had eased. When he left the hospital three days later, his potassium and heart rate were normal and his kidneys nearly so. He would have to keep the tube in his bladder until his prostate could be removed.

I was out of town that first day and had to follow my patient's progress by telephone. When I heard that the prostate was the cause of the life-threatening bradycardia, I felt as if I had been punched in the chest. This was something I should have caught and didn't. An internist's job is to diagnose and treat acute illness and screen for and prevent additional disease. I joke with the residents I teach that it is our responsibility to keep our patients healthy and out of the hospital. If so, I had failed.

Screening for disease has two parts: usually a physical exam and what is known as a review of systems, a set of questions used to elicit symptoms of a disease the patient is at risk for. This patient, with his high blood pressure, high cholesterol, and stroke, would be at risk for a heart attack, another stroke, and, like many men his age, prostate problems. I should have asked about these at every visit and once a year done a rectal exam to assess prostate size and look for cancer. From reviewing the patient's chart, it appeared I had limited my attention and my exam to his immediate problems—overlooking some of the other risks he faced.

I had asked him if he had problems urinating, and he had said no. I don't think he was lying; maybe he didn't. I think it's possible he assumed that his bathroom difficulty was just one more skill stolen from him by his stroke. So much of the damage from that cerebral vascular accident was clearly visible and public. I suspect he felt that this disability, at least, could remain private.

And when he didn't acknowledge any difficulties, I was happy to allow our visits to focus on getting his blood pressure and cholesterol under control, educating him on his medical problems, managing his meds, and arranging his transportation and rehab. Everything else I treated as a long-term goal, to be attended to once these very pressing short-term needs were managed. Understandable perhaps, but it almost killed him. Practicing medicine is a balancing act—weighing immediate and long-term good. His case was a vivid reminder of what can happen when that balance is lost.

I didn't visit my patient in the hospital. Normally I would have, but I was worried that he would be as angry with me as I was with myself. I saw him the following week. "I'm so sorry," I started. He smiled his magnificent smile and squeezed my hand. "No matter," he said, his words still slurred but back to their normal rhythm. He reached into his pocket, produced a few of his pecans from North Carolina, and offered them to me. I took them gratefully. Perhaps I could be forgiven.

ACKNOWLEDGMENTS

This book quite literally originated in the pages of *The New York Times Magazine*. It was born out of a question asked by editor Paul Tough back when I was still a doctor in training: What can doctors write? These columns are a version of the story physicians write every day. I'm grateful that Paul and then editor in chief Gerry Marzorati believed these were stories worth telling. I'm indebted to Dan Zalewski for persuading me that I could be the doctor who told them. And to the many editors who made these columns better. Big thank-yous to Joel Lovell, Catherine Saint Louis, and Katherine Bouton. Ilena Silverman has stuck with me for over a decade. Her wisdom and curiosity have kept the focus of these stories where it belongs: on the patient and on the diagnostic process, as well as on the disease. Thank you also to Jake Silverstein for his vision of what the magazine could be, and for his willingness to make a place for me in it. And to Mary Silverstein, who was a fundamental part of that process, many, many thanks.

This column could never have survived without the help and support of my friends and colleagues at Yale. Thank you, Ralph Horwitz, for your early and enthusiastic support. Thanks also to Gary Desir and Patrick O'Connor, who believe that the doctor's voice belongs in public discourse—even when talking about the complicated and not-always-rational process of diagnosis. I am inspired and supported by all the terrific doctor-writers I work with: Vincent Quagliarello, Marjorie Rosenthal, Randi Hutter Epstein, and especially Anna Reisman. Thanks also to the doctors who teach me about medicine every day: John Moriarty, Steve Huot, Julie Rosenbaum, Donna Windish, Auguste Fortin, Tracy Rabin, Joe Ross, Cary Gross, and the rest of the internal medicine faculty at Yale. I am indebted to all the fantastic residents and medical students I work with in the clinic and in the wards, who ask the questions that keep me thinking and learning. Thanks also to Andre Sofair, Tom Duffy, and David Podell, who continue to teach me about the diagnostic process. Thanks also to the many doctors who contributed their experiences to my columns, examining their thinking and their notes to reveal the complexities of the diagnostic process and how it goes right—and occasionally wrong. I learn from you every day.

The process of putting together a book is complicated and, really, so much harder than I ever expected. This book would not have been possible without the dedication and persistence of my friend and agent Gail Ross, who believed in me long before I wrote a single line. Her regular, one-line emails about the projects in my head are the one reason any of them ever appeared on a page. I'm grateful to Amanda Cook and her amazing team at Crown, who really went all out for this book. And to my editor, Zachary Phillips, who helped me say what I wanted to say, only better. Closer to home, many thanks to my part-time research assistant and full-time medical student, Fatima Mirza, who helped me put this collection together in her spare time between studying for tests and dissecting cadavers.

I'm also grateful to Scott Rudin, who saw that the real drama in medicine can be even more powerful than fiction. The series *Diagnosis* couldn't have been possible without his vision, along with the creativity of Jonathan Chin, Alex Braverman, Alyse Walsh, Peter Morgan at Lightbox, and Kate Townsend at Netflix.

Almost everything I know about writing I learned from my husband, Jack. A great writer and the best, most supportive, thoughtful, and loving partner I could imagine. You have collaborated with me on all my best works. I owe you big. My children, Tarpley and Yonce, have been a constant source of inspiration to me. I love how they were able to love and support me while they were in the midst of this act of self-creation we call growing up. I am so lucky.

But most of all, I am grateful to the many patients who, over the years, have been willing to share their stories with me, reliving some of the most difficult times of their lives and the lives of those they love, so that others could learn from their experiences. For that reason, and so many more, this collection is dedicated to them.

ABOUT THE AUTHOR

LISA SANDERS, MD, is an internist on the faculty of the Yale University School of Medicine, writes the monthly column Diagnosis for *The New York Times Magazine*, and was the inspiration and adviser for Fox TV's *House, M.D.* She lives in New Haven, Connecticut.

ABOUT THE TYPE

The text of this book was set in Janson, a typeface designed about 1690 by Nicholas Kis (1650–1702), a Hungarian living in Amsterdam, and for many years mistakenly attributed to the Dutch printer Anton Janson. In 1919, the matrices became the property of the Stempel Foundry in Frankfurt. It is an old-style book face of excellent clarity and sharpness. Janson serifs are concave and splayed; the contrast between thick and thin strokes is marked.